THE REPROGRAPHIC LIBRARY

Electrostatics in Reprography

Electrostatics in Reprography

by

W. A. Cook A.M.I.R.T.

Focal Press
London & New York

ISBN 0 240 50699 5

Published in co-operation with
the Institute of Reprographic Technology

PRINTED AND BOUND IN GREAT BRITAIN BY
A. Wheaton & Co., Exeter, Devon

CONTENTS

Preface 7

1. *Applications and Limits* 9

What electrostatic printing is. What electrostatic printing can do. Electrostatic printing machines. What electrostatic printing cannot do.

2. *Development of Electrostatic Processes* 17

First steps in electrophotography. Carlson's unorthodox approach. Producing the first image. Birth of xerography. Commercial developments.

3. *How Electrostatic Processes Work* 27

Electrostatic and electrographic. Definition of xerography. Commercial electrostatic processes. Additives for photo-receptors. Photo-receptor carriers. Producing the image. Developing and fixing the image. Basic processes and variations. Electrographic processes.

4. *How the Photo-receptor Works* 43

Electrical qualities. Atoms and their electrons. Charging the photo-receptor. Discharging the photo-receptor—theory. Photo-receptor characteristics. Zinc-oxide photo-receptor. Elfasol photo-receptor. Furnium and Xerox photo-receptor. Cadmium sulphide laminated photo-receptor.

5. *Imaging the Photo-receptor* 56

Function of the lens. Additional function of the lens. Control of light intensity. Use of mirrors for right-reading images. Path of image to photo-receptor. Illumination of original. Lens-less equipment. Possible developments.

6. *Developing the Image* 68

Toner characteristics. Charging the toner. Cascade development. Electrophoresis or liquid dispersion development. Magnetic and fur brush development. Aerosol or powder cloud development. Reversal development. Development control techniques.

7. *Preparing and Handling Originals* 80

Choice of system and equipment. First and second generation copies. Examination of originals. Unalterable characteristics. Controllable characteristics. Preparation of originals. Manually-operated machines.

8. *Print-out Media* 89

What paper is made from. How paper is made. Storage and use of paper. Paper grain. Desirable characteristics for uncoated stationery. Sensitized papers. Characteristics of zinc oxide papers. Limitations of machine design. Controlling the output.

9. *Review of Current Equipment* 99

Preliminary survey. Review of equipment. Cost per copy calculations. "Best method" considerations. Comparison of cost figures.

10. *Process and Machine Projections* 127

Colour problems. Line and tone colour. Liquid-development colour methods. Dry cascade methods. Output speeds. Canon NP systems. Elements of PIP methods. Competitive processes. Long-distance xerography. Storage and retrieval methods.

Appendix 143

Glossary 152

PREFACE

That an enormous amount of scientific information and data exists to explain what static electricity is, how it is created, where and when it is used, is beyond question. Much of the information is contained, however, in monographs or textbooks written for those with scientific training.

This book bridges the gap between the research scientists and those who wish to use their findings in profitable, useful or necessary enterprises. It has been written in the idiom of the day from the point of view of the business manager, commercial student, the administrator and the technicians who are concerned with the production and reproduction of the many documents that keep commerce moving and without which the business of the nation would come to a halt.

From the many, diverse and often relatively inaccessible sources, only the information that is necessary for the non-scientific reader to understand and use static electricity has been concentrated in this volume. For those who wish to delve into the mathematics and the science of electrostatics, one or other of the reference books listed under the heading "Recommended Reading" is suggested.

The author acknowledges his deep appreciation of the help given him by his colleagues in the industry and the members of the Institute of Reprographic Technology, who vetted the manuscript to ensure that it was unbiased and accurate. His grateful thanks are also extended to the companies which sponsor the equipment and processes described in this book, The Radio Corporation of America and the directors and staff of Rank Xerox Limited for providing him with the information which has enabled this book to be written.

Windsor 1970 *W. A. Cook*

CHAPTER I

APPLICATIONS AND LIMITS

Printing, in the wider sense, involves making marks or figures on surfaces, particularly paperlike surfaces. Methods of printing are, in effect, methods of taking ink from a reservoir, breaking it up into the form of letters, figures, diagrams and pictures and then transferring it to paper or some other similar media. All the traditional processes use pressure somewhere in the system, usually at the point where the required ink pattern makes contact with the media which is to be printed. This is even true of the typewriter, where the ribbon acts as the ink reservoir.

It is easy to understand where the pressure is used in the letterpress and offset litho methods of printing. These machines have impression cylinders the pressure of which can be adjusted to suit the media being printed. In the stencil system, the ink is forced through holes in the stencil representing the image to be duplicated. In the spirit system, the ink reservoir is in the letter or figure itself, which is moistened with spirit and lightly pressed on to the copy paper. The intaglio or gravure process uses pressure applied by rollers of varying degrees of hardness to achieve transfer. The doctor or scraper blade is very finely set in order to remove surplus ink from the non-image areas. Without pressure, none of these processes could produce an image.

What electrostatic printing is

Electrostatic printing methods are radically different. The "ink" is a very fine thermoplastic pigmented powder made to drift into place under the influence of electrostatic or electric charges that represent the desired image. The process is neither chemical nor photographic in the sense that the image is not developed from the latent chemicals previously impregnated into the paper or other media being photographed. It is nevertheless a printing process within the ordinary meaning of the term. It could perhaps be described as a pressureless printing process.

Electrostatic printing is not a gimmicky term for a process hitherto known by another name. It is a term which describes techniques belonging exclusively to the twentieth century. The key invention which reoriented the scientific approach towards using electricity to create viewable images came in 1938. The invention was patented in 1939 under the heading "electrophotography". The war years delayed its development and it was not until 1950 that commercially usable hardware appeared. It is from this mid-point in the twentieth century that electrostatic printing techniques should be dated.

It is an entirely new electrical process for producing viewable images of all sorts on paper and similar media of all kinds including the better quality ordinary office stationery which is in everyday use. Not only can the process print in its own right, but it can be associated with other printing processes. When this is done these processes have wider and more useful applications and become more convenient, efficient and economic.

What electrostatic printing can do

Electrostatic printing machines have a variety of functions, limited only when they are designed with specific tasks in mind. Machines designed to image specially coated papers, for example, cannot make copies on plain uncoated paper. Machines designed to image one side of the copy paper cannot image both sides, and so on. In general, however, they can:

1. Operate direct from the original without needing a special intermediate master to be created. They are fully automated and are press button controlled. A master is that which any particular process needs from which it generates further copies.
2. Make reproductions of original documents in large or small numbers, slowly or quickly at various print quality levels on flat surfaced paper or similar materials. Some collate the copies automatically as they print at speeds up to a copy a second.
3. Make the copies either on specially coated proprietary papers or on ordinary paper in everyday office use.
4. Make new originals by adding, deleting, substituting and/or separating information and data from existing documents and print the new copies in one operation.
5. Make intermediate masters for other reprographic processes

such as offset litho and diazo. With some machines, the special paper used is also suitable for offset litho duplicating purposes. Machines that use ordinary paper can be loaded with the appropriate master media.

6. Offer, in a few cases, enlargement and/or reduction facilities. Most machines, however, work on a "same size to original" basis.
7. Operate from multicoloured originals but cannot yet separate or reproduce the original colours. The colours of the originals are rendered as shades of grey; black is rendered as black; but white is sometimes a pale grey. The more expensive the copy, the better the quality.
8. Print, in some cases, 16 mm or 35 mm microfilm in roll form or in unitized aperture card form on to plain or coated papers.
9. Make copies from copies made by the same process.

In terms of recognized applications and with a greater or lesser degree of efficiency, convenience and economy, these machines can copy and/or duplicate the following:—

Invoices	Data Computer Output	Training Notes
Letters	Price Lists	Technical Publications
Diagrams	Parts Lists	Agreements
Press Cuttings	Extracts from books	Index cards

In fact, they can copy anything from a page in a thick book to a wafer thin airmail letter or a rigid strip index panel.

The more modern types of machine can create new documents from originals without any of the information and data on the original documents having to be transcribed by hand. Like computers, electrostatic printing machines have acquired their own particular collection of peripheral and ancillary equipment. These are attachments which can be fitted to the machine and which enable the operator to add, delete or substitute items of data and information contained in the original document.

These attachments consist of masks, overlays and templates and the general name given to their use is *templatry applications.* There are templates for titling and/or routing documents as they are printed and for registering originals in one place while they are being duplicated. This is useful when copying library cards. Small library and index cards can be copied two or four up. Thus one machine revolution can handle up to four originals at one time.

Masks and overlays can be used in combination in a number of ways. A departmental requisition, for example, can be converted into a supplier's order, a goods-in advice, an expedite copy, an invoice payment authority and as many other statistical and managerial copies as may be required.

Similarly, sales order control systems can be used where the originally created internal order can be used to produce an advice note, an invoice, warehouse dispatch note, delivery note, field sales advice, accounting, marketing and statistical copies.

These special uses are known as *system applications*. They illustrate how modern electrostatic printers can do all the work hitherto done by continuous stationery typing, offset, stencil or spirit duplicating methods. Whether it is convenient, economical or advisable to use these machines in this way depends on form design and various other considerations.

Some machines are designed for only one application, such as The Rank Xerox CFP, which reduces computer tabulations to A4 size, overlays the appropriate headings and automatically collates the copies.

Electrostatic printing machines

Machines using electrostatic printing processes fall into three broad categories: rotary—often referred to as desk-top machines, flat-bed or console models, and electrographic machines.

A rotary machine in the electrostatic printing context is one where the original disappears inside the machine where it remains until the copying or duplicating cycle is complete, and then is returned. A flat-bed machine is one where the original rests on a flat platen under a cover, or a template, mask or overlay until the copying or duplicating cycle is complete.

These radically different designs affect the extent and versatility of the application.

The rotary machines are restricted to printing from flexible originals which are also relatively thin. These machines can be used for systems applications provided that the masks, templates and overlays are also thin and flexible. Obviously, these machines cannot make copies of pages in bound books, nor could they be used for systems involving rigid strip index panels. They are however, often smaller than the flat-bed machines. Rotary machines are often found on a secretary's side table and are used to eliminate the need to take carbon copies when more than two or three are needed. They are very useful for copying incoming correspondence so that a manager can initiate simultaneous action within his staff by noting the action on the original and having the copies distributed direct from his office, complete with his directions as to the action to be taken by the recipients.

The flat-bed machines on the other hand, can use as an original, rigid strip index panels, thick books as well as very thin papers, ordinary files and folders. The applications include all those mentioned previously plus some others associated with the wider range of originals mentioned above.

The use of strip index or Quicklista systems with modern fast automated copiers eliminates the use of standing type for electoral rolls and the retyping of lists of all sorts. Similar systems can be used for keeping up to date and the immediate printing and simultaneous printing and collation of stock lists, parts lists, price lists, schedules of loose tools, movable asset inventories and nominal rolls of university students as they progress from class to class or one faculty to another. The creation of credit transfer advice slips for standing order and traders credit transfer transactions direct from strip index panels is another application which has become possible by the use of electrostatic copiers and duplicators. This is an example of how entirely new documents can be created from entirely different original records. The use of strip index panels in conjunction with these new copiers is limited only by the imagination of the user.

With the right templates, flat-bed machines can be used in conjunction with ledgerless accounting paperwork, routines, the marking of copies during production with addressee destinations and the production of lists from file cards, library cards or slips of paper by using shingling methods. "Shingling" means that the slips of paper or card are laid out in such a way that the edges which carry data or information overlap and therefore show up one beneath another.

An example of the shingling application arises from the use of car ferry loading authorization slips issued to vehicle owners and drivers by the shipping company, its agents or the motoring organizations. These slips, which carry the name of the driver, make and registration mark of the car, and other information, have to be listed for customs, insurance and shipping purposes. The ship cannot sail without the lists. The vital information to be listed is shown along the top edge of the slip. Thus, when they are inserted into the pockets of the shingling template, only the top edge that carries the requisite information shows above the slip in the next pocket below. The template containing them, has only to be placed on a flat-bed copier for the list of vehicles which passed through the control point to be listed, without any of the original information having to be transcribed in any way. Library lists by author or title could similarly be produced from suitably designed library cards placed in a shingling template.

What electrostatic printing cannot do

Over very long print runs from one original, electrostatic printing machines are not so economical as the offset and stencil methods of

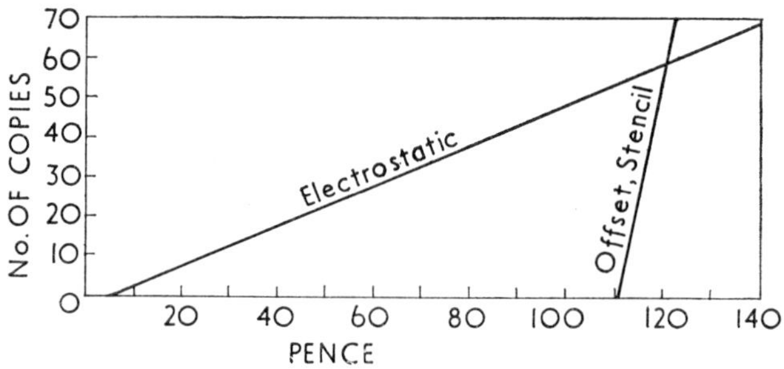

Fig. 1. Notional cost comparison between electrostatic and conventional duplicating processes.

reproduction. Fig. 1 gives a break-even point comparison. The precise point at which any one user switches from one process to another varies, because costs vary in different organizations.

Print quality from electrostatic machines is sometimes better than that produced by some of the other processes, but if the machines are maladjusted or badly maintained the print quality is below the standards expected of the traditional process. In the last few years, enormous strides have been made in improving electrostatic printing machine line quality. The latest machines on the market really do produce copies of typescript which are indistinguishable from the original. The white areas are, in fact, background free and the images are as black as the original.

These new machines do not yet seem to be able to handle half tone illustrations and solid black areas as well as some other processes. This is particularly true when half tone illustrations and solid black areas appear as part of the original. Some of these new machines can handle solid black areas better than others, but in doing so tend to sacrifice resolution and therefore half tone quality veers towards "soot and whitewash" type illustrations.

Generally speaking, electrostatic printing processes cannot match the overall quality of print produced by highly professional offset operators working with photographically produced offset litho masters. It is a matter of opinion whether an electrostatically printed copy is of better quality than a copy produced by the stencil process. In some cases it is and in other cases it is not.

Electrostatic printing processes have not appeared in offices as machines which can separate the colours out of a multi-coloured

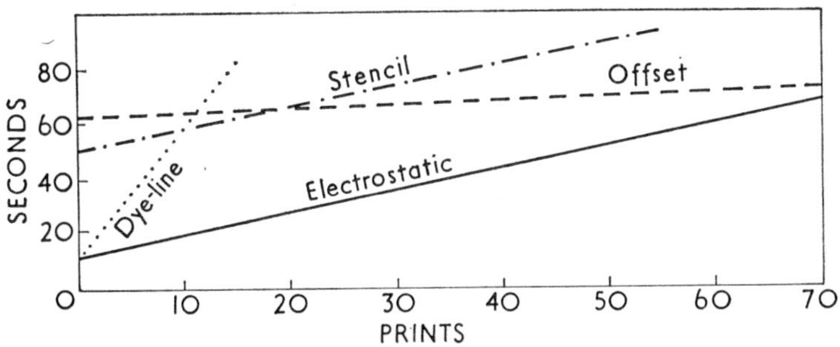

Fig. 2. Notional time delay comparison between first and subsequent copies produced by various reprographic methods.

original, nor can they print two different colours on one sheet of paper . . . yet! This qualification is made because the Electrofax method has been used to produce two or three colour prints mainly for military purposes but, as yet, it cannot be considered to be a method which is suitable for the executive suite. It is more suited to the laboratory or to the print room.

Speed of output of one machine is not a relevant basis of comparison with other machines. It is effective overall speed in relation to the time that the original leaves the originator to the time that the copies return to the originator that really matters. In the case of short print runs from one original, the electrostatic printing machines can be expected to score over the traditional processes, but where long print runs are involved, then one or other of the traditional methods will prove quicker. Fig. 2 illustrates this.

If the traditional methods are operated centrally there is transit delay time while the original travels from the originator to the central print room and the delay time associated with the making of the master and the return of the copies to the originator during which the electrostatic printing machine located in the originator's office can make copies. On balance, machine operating speed of itself means very little and in any case, the fastest of the electrostatic printing machines now operating produce copies at the rate of one every second. Some of the conventional duplicators may need more or less time to be spent on the make-ready procedures, the time for which was included in the figures used to produce the diagram in Fig. 2. The actual figures involved may vary from company to company and machine to machine but whatever is the make-ready time, the electrostatic printing machine operating at the point of need and origin will make some copies during the make-ready time.

CHAPTER II

DEVELOPMENT OF ELECTROSTATIC PROCESSES

Electrostatic methods were developed as a result of the inadequacies of the reprographic processes extant towards the end of the nineteenth century. The first impact of the demand for improved methods of producing visual communications is noticed in the industrial sphere for the reproduction of architects' and engineers' drawings. Those engaged in business and commerce seemed to be content with the methods they were using.

At the turn of the century, photography had had about sixty years in which to develop. It solved the problem of reproducing engineering and architects' drawings by a method that became known as the blue print or ferro-prussiate process. It was cumbersome, slow and expensive, but it largely superseded hand tracing. Cheaper copies were provided by the true to scale process or Ordoverax Veleography, which is lithographic in principle.

The visual communications requirements in industry continued to be in advance of the technology available. Another step forward was made when the dyeline or diazo process, scientifically noted in 1885, was patented in 1921 by Ozalid Limited. Although it had the limitation of requiring originals to be translucent it cheapened and made more convenient the reproduction of engineering and other drawings.

Photographic methods were introduced into offices as a result of the need to reproduce sketches, diagrams and pictures. At first, direct and later reflex photographic methods were used. In order to solve the problems these created and in order to overcome some of their operational limitations, the gelatine transfer, diffusion transfer and direct positive methods were successfully introduced. It was photography that helped to create the office photo-copying market that is now largely dominated by electrostatic processes.

The use of electrostatics as a reprographic tool belongs quite clearly to the twentieth century. Up to the first world war, electrostatic studies were left to dilettante intellectuals. Electrostatic forces were, however, put to some useful purposes. Metal tiles were imaged with powders deposited in attractive patterns under the influence of electrostatic fields of force and then burnt into the surface when the tiles were furnace fired. Piled surfaces were produced by allowing elongated fibres to be driven into an adhesively coated base material under the influence of electrostatic fields.

Electrostatics was used to produce ornamental pigmented patterns on a web of paper by wet methods. Then again, ornamental powder patterns were produced on cloth by methods involving the use of stencilled designs and electrostatics. None of these promised to make a fortune for either the inventor or his sponsors.

First steps in electrophotography

Between 1920 and 1939, R. Pohl in Gottingen in Germany carried out a systematic study in photo-conductors. He proved scientifically that the effect of light on a photo-conductor increases the motion of electrons and in consequence the electrical resistance of the material is reduced. He further discovered that the conductivity of a photo-conductor depends upon its purity, and also that resistivity decreases as the temperature increases.

In 1922 a patent was issued to Von Bronk. This was probably the first purposeful attempt to devise an electrophotographic process. In this process an electrosensitive paper is sandwiched between a photo-conductive layer and a conductive sheet. A voltage is applied across the sandwich. The current flow is held in check under the influence of light. The current flows only in the illuminated areas and burns an image into the electro-sensitive paper. At best, this might be considered to be a forerunner of the Videograph, but the process can hardly be called electrostatic. Similar methods were described by Hana and later patents employing similar methods by Goldmann, Schmieschek and Klutke were registered.

It is often thought that Paul Selenyi between 1920 and 1930 developed electrophotographic recording methods. Strictly speaking they are electrographic rather than electrophotographic. The inventions he sponsored did not involve photographic images. He produced images which were written by an electron or ion beam, which created electric charge patterns on insulating surfaces by scanning. The electrical image was made visible by dusting with powder. This was the forerunner of the A. B. Dick Videograph and the Xerox Mobile Printer.

In 1931, William C. Huebner patented an electric printing process in which electrostatic fields were used to attract ink to the paper from the printing plate. Although some 20 or so US Patents have been issued subsequently to Huebner, W. J. Hooper and the Goss Printing Press, covering variations of the process, it has not yet emerged as a popular commercial process.

Carlson's unorthodox approach

In 1930, a man called Chester F. Carlson got his B.Sc. This academic accomplishment was achieved against a background of great personal adversity. His father was bedridden for most of his life and his mother died when he was 17. His formal education was interrupted from time to time with spells of work in order to pay his way. His B.Sc. led to a job in the Bell Telephone Laboratories in New York, so he had to leave his father in California and support him from New York.

Because he did not enjoy bench work at Bell Labs, he transferred to the patent department of P. R. Mallory & Co. He liked the combination of science with legal work. As he had a scientific background he decided to acquire legal knowledge through evening classes by attending New York Law School. He became a patent attorney by 1934.

Similar work by de Meulenaere. It was a brief description of Paul Selenyi's work on the electrostatic formation of images by facsimile scanning that appeared in New York in 1937, that spurred Carlson on to take out a preliminary patent. De Meulenaere nearly anticipated him in 1932 by taking out a Belgian patent. His idea was to expose a selenium surface coated on tin foil to an image in a camera. The selenium was not charged prior to exposure. After exposure a subsidiary conductive surface pre-coated with a powder was brought near the imaged selenium surface, and the two connected up through a Leyden jar. He hoped that the Leyden jar charge would be patterned by the selenium as its charge was discharged and so remove powder from the subsidiary surface in proportion to the light and dark shades induced on the selenium surface. The subsidiary image thus created electrostatically could be fixed by a shellac spray. This invention was never developed into a practical process. Had de Meulenaere pursued his theme, he would probably have evolved the idea which is the key element in Carlson's invention—the photo-electric formation of latent-electrostatic images.

Reversing the procedure. From the foregoing account of the scientific developments, one cannot avoid the conclusion that the orthodox scientists of the day were trying to make electricity write and copy by positively controlling the electrons that do the job. Orthodox science was getting into a tighter and tighter scientific straightjacket with the technology then available, the more it experimented along orthodox lines. They had all been trying to put down electrostatic charges where they were required to make the image. It took the ingenious mind of Carlson to think against the main stream. He must have said to himself either consciously or sub-consciously, "Instead of putting down charges where we want them, why not wipe the charges away from where we don't want them?"

At a time span of 30 years, at a space displacement of 3,000 miles and working through published literature, the student can only conjecture what went on in an inventor's mind. Up to the start of Carlson's investigations of electrostatic image formation on photo-conductive layers in 1939, all the academic thinking had not produced a viable electrostatic method of reprinting the written word. The converse to all the methods so far tried was the only course left. It had to work, and it did. Carlson's determination to make his idea work was probably a greater driving force than that which motivates the academic scientist. Carlson had had to develop a determination to overcome obstacles at a very early stage.

Not only as a patent attorney did he experience frustration and dissatisfaction with the then existing methods of copying documents and drawings and thus have a motive to invent a new process. His motivation probably also arose from those early boyhood experiences in helping to produce a high-school magazine for amateur scientists. It was then that he came face to face with the inconvenience, expense and complexity of the reprographic technology in being at that time. In his business capacity years later, he found that there had not been many fundamental changes in the methods used.

Between 1934 and 1937 when he set out to invent a new process, Carlson reviewed scientific literature in the New York Library and began to sort out ideas as to how photo-conductivity and electrostatics could be used with a photographic plate so that electric currents could be controlled by a light pattern or image.

He engaged a refugee physicist named Otto Kornei in 1938 to work with him in his New York apartment to prove his idea. The smelliness of his experiments with sulphur caused complaints and he had to transfer his activities to another area.

Producing the first image

The first electro-photographic plate consisted of a layer of sulphur mounted on a zinc plate. Having been fused, the sulphur was spread on the zinc and allowed to cool and harden before being rubbed with a handkerchief in order to produce a charge. The operation was carried out in the dark.

The original he chose for his first image, was to be an alpha-numerical one. It was this type of reproduction that was required and it was this type of image that had stumped his predecessors. He wrote the date, American fashion, with the month first, followed by the day and the year. The alpha part of his image was represented by the

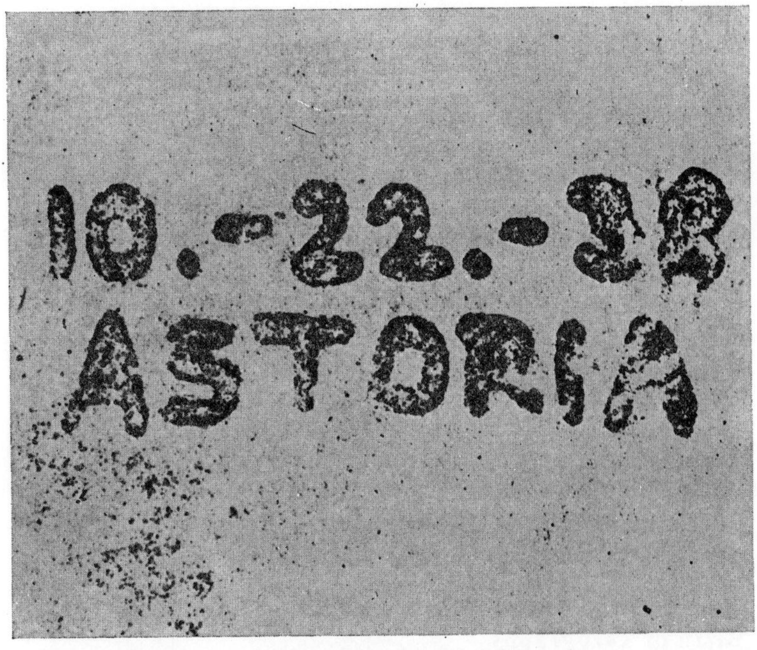

Fig. 3. The first electrostatic alpha/numerical image was created on 22nd October, 1938, in the Astoria district of New York by Chester F. Carlson to prove the viability of his invention of the electro-photographic process.

district in which his lab-kitchen over a bar and grill was situated. So his first image was to be:

10 — 22 — 38
ASTORIA

He wrote this legend on a glass slide and contact exposed it, the wrong way round, to the charged sulphur surface. Having switched off his light source, he poured over the image a dyed lycopodium powder which stuck to the sulphur in the image areas. A piece of wax-coated paper was pressed against the powdered image with a roller. When it was peeled off the plate, the first electrophotographic image on plain non-photographic paper was made.

Reprographically the quality was very poor, but it proved the process. All that was necessary was refinement and improvement.

Patents for process and apparatus. Having reduced the theory to practice, Carlson sought to protect his idea. He applied for and obtained his first patent. The patent protection covers the process called electrophotography. Its date is 4 April 1939 and illustrations from it are reproduced in Fig. 4. In effect, it is a definition of terms in that the patent explains what is meant by the term. The date is important for it authoritatively and officially refutes any claims that sometimes are made by sponsors of processes derived from Carlson's invention that the derivatives came first.

Nineteen months later, in November 1940, Carlson obtained another patent, this time for apparatus using the process. It is interesting to note the rotary concept enabling all the stages of the process to take place at stations fixed at positions around the periphery of a rotating drum. This concept did not manifest itself in a commercial copying machine until about 1960 in the USA and 1961 in the UK when the RX 914 copier was launched.

The model that Carlson had made, on which his patent was based, worked successfully but frequently broke down owing to minor mechanical defects. By that time the mechanics normally available to make repairs were fully occupied on war work so the model was not developed into a prototype machine. It is shown in Fig. 5.

Birth of xerography

With his process and machine patents secure in his name, Carlson in 1940 began in earnest to try to capitalize on his invention by enlisting commercial support for its exploitation. Some twenty companies turned it down. Even the National Inventors Council in the USA and the US Signal Corps were disinterested.

Ultimately in 1944, the Battelle Memorial Institute of Columbus, Ohio, agreed to carry out further research into his process. Under

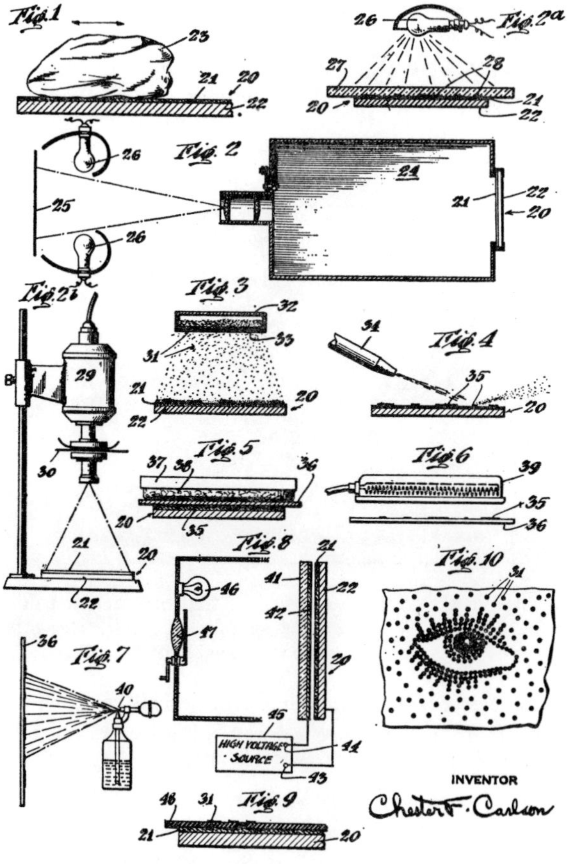

Fig. 4. Illustrative extract from Chester F. Carlson's first patent of the electrophotographic process.

this agreement the Institute undertook to do development work. In 1947 the Haloid Company (now Xerox Corporation), then a small photographic supplies manufacturer, became interested in Battelle's work through the vigilance of Dr. John Dessauer who read an article by Nicholas Langer in *Radio News*. Haloid supported Battelle's development with the result that Bixby discovered that amorphous or vitreous selenium deposited in a high vacuum on brass or aluminium became a photo-conductor with a much greater light sensitivity than the sulphur and anthracene layers used previously by Carlson.

Walkup invented the screen controlled corona for charging Bixby's plate, while Walkup and Wise between them produced developer mixture of pigmented powder mixed with a carrier bead. This was found to be more efficient than powder alone. Finally, Schaffert

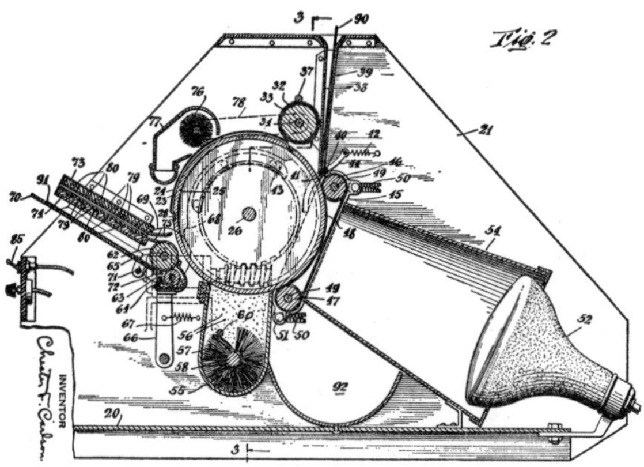

Fig. 5. Illustrative extract from Chester F. Carlson's first patent for electro-photographic machines. With the exception of a lens, all the elements of modern machines were envisaged in 1940.

invented the electrostatic transfer method. This released the process as a whole from dependence on the adhesive method of transfer used up to that time.

In 1948 on October 22—ten years to the day after the invention—the process was demonstrated by Haloid and Battelle to the Optical Society of America in Detroit. At that time it was given the name "xerography" meaning dry writing from the Greek "xeros" for dry and "graphos" for writing.

Commercial developments

In 1950 the first commercial copier was launched on the US market. It was a manually operated unit. A modern European version was launched in the UK in 1957 and there was no other electrostatic copier on the market until 1961. The original unit was sold and rented primarily as a mastermaker for the small offset duplicators and printers.

The emergence of this first commercial electrostatic copier encouraged a large growth in research and invention. Using Carlson's idea to use photo-conductivity to disperse electrostatic charges in the non-image areas, researchers investigated other types of organic and inorganic powder in binders to form photo-conductive surfaces.

In 1954, Young and Greig of the Radio Corporation of America were two such workers who announced an Electrofax process using a photo-conductive surface of zinc oxide pigment in a resin binder bonded into a paper backing.

The Electrofax process "know-how" was offered by RCA for development under licence. The Charles Bruning Co., American Photocopy Equipment Company, Keuffel and Esser, the SCM Corporation and others, both in the USA and Europe, took up RCA's offer. Now there are some 40 makes of machines using the zinc oxide based process which claim about a third of the electrostatic market.

In 1958, the Minnesota Mining and Manufacturing Co. introduced a process using the phenomena of persistent conductivity discovered by H. Kallman in the USA and V. M. Fridkin in the USSR. The Filmac microfilm reader printer uses the Kallman process.

Whereas with both the selenium and zinc oxide photo-receptors cascade and magnetic brush development methods were used respectively, a liquid method for charging and carrying the powdered pigment to the residual static charges on zinc oxide coated paper was evolved by Mayer and associates at Horizons Inc., in the USA in 1957 and by Metcalfe and associates in the Defence Standard Laboratory in Australia in 1955.

It was in the 1960s that business management began to feel the pinch exerted by the chemically based sensitized paper process. The succession of processes from photography through the direct, reflex, the various transfer methods to the direct positive method, showed that management demanded something better or more convenient or cheaper. The diazo process entered the field to solve some problems and create others. Then came the heat processes. Again some problems were solved and new ones created. This was followed by the diathermic or dual spectrum process, again the same pattern, all problems arising from the use of sensitized papers. These developments were terminated at the introduction of adherography and so the process which sprung from chemistry reached, the long way round, what management was looking for—a process for copying images on plain unsensitized paper.

CHAPTER III

HOW ELECTROSTATIC PROCESSES WORK

The electrostatic process is one of two groups of powder printing processes. The dry powdered ink that makes the image visible is attracted electrostatically to the media on which the image is to appear. No physical force is used initially to produce the image. Physical force may or may not be used to control the flow of paper through the machine or to improve copy quality, or make the image more permanent; but physical force or pressure is not essential to the image-creating process itself. No pressure is applied at any stage.

Electrostatic and electrographic

The other powder printing process is known as the electrographic process. The two processes are easily distinguished. In the electrostatic process, static electricity is first spread all over the surface to be imaged and subsequently dispersed from the non-image areas. Thus the static charges remain in the wanted image areas. These charges are used to form the visual communications that appear when these remaining electrostatic areas are powdered with dry "ink" more commonly referred to as "toner" to distinguish it from the conventional wet or greasy variety. It is, in fact, thermosetting pigmented powder.

In the electrographic method, electric charges rather than electrostatic charges are positively positioned where they are required on the media that is to be imaged.

Chester F. Carlson was the inventor and patentee of the first group of processes, whereas Paul Selenyi did most of the preliminary work from which the second group were later developed jointly by the Stanford Institute of Menlo Park, California and The A. B. Dick Company of Chicago, Illinois.

The electrostatic processes are the most widely used, with 70,000 or more machines in operation in various parts of the world. Comparatively few machines use electrographic methods. Electrographic machines are large industrial reprographic units needing highly specialized applications and very large copy volume to be economic. They are unlikely to be suitable for, or to be found in, the ordinary office. Like the multi-colour Electrofax electrostatic process (see page 15) these electrographic machines are more suited to the laboratory or to the print room than to the executive suite of offices.

Definition of xerography

The electrostatic group comprises the original process patented as electrophotography and which, some ten years later, was given the name xerography (which is not a trade mark or trade name) by a totally outside, independent and professional body called The Optical Society of America in Detroit on October 22, 1948. An up to date definition of this word is as follows:—

> *XEROGRAPHY:* a process for making copies of printed, written or pictorial material from film or paper, in which areas on a photo-receptor corresponding to those areas on the original that are to be reproduced are sensitized with a charge of static electricity so that, when powdered with a coloured resin carrying an opposite charge, only the charged areas retain the resin, which is then fused to the photo-receptor or transferred to plain paper and fused to make it permanent.

R. M. Shaffert MS, PhD, in his book *Electrophotography*, page 21, reports as follows:—

> The IRE (now IEEE) committee on electrostatographic devices has been working toward the standardization of terms and definitions involving electrostatic images. This committee has defined xerography as "that branch of electrostatic electrophotography which employs a photoconductive insulating medium to form, with the aid of infrared, visible or ultra-violet radiation, latent electrostatic charge patterns for producing a viewable record."

Taken together these two independent definitions indicate that the term xerography covers all processes in which an electrostatic image exists at some point in the process. Thus the RCA Electrofax process and the Kalle process should both be classed as electrostatic methods for producing images within the general xerographic process.

The diagrammatic summary of the three interrelated methods given in Fig. 8 follows the general understanding that the term xerography embraces all three methods.

Commercial electrostatic processes

For easy identification of the three methods that comprise the electrostatic group of processes, the names of the companies or corporations which sponsored the method or the names given by these companies to the methods that they sponsored, are used to refer to them:

Xerox Method. Xerox is a trademark owned by Xerox Corporation.

Electrofax Method. Electrofax is used by RCA to describe their patented process, but the word is not registered as a trademark; it is associated with the RCA process by usage.

Elfasol Method. Elfasol is used to denote the method that the Kalle Co. of Germany sponsor.

The Xerox method is associated mainly with the imaging of ordinary non-proprietary office stationery and the production of ordinary non-sensitized paper and metal masters and for the production of diazo bond masters for the diazo or dyeline process.

The Electrofax method is associated with the imaging of specially coated papers which can also be used as short run paper offset litho printing masters.

Both processes are used in automated office copiers and duplicators. The Xerox process can additionally be applied in manually operated equipment which provides enlarging and reducing facilities and can be used for producing metal offset printing masters as well as for making paper offset masters and transparent sheets of various plastic material for the diazo process, the making of lantern slides and similar applications. It also has some other specialized uses and has been on the UK market for about twelve years and since 1950 in the USA.

The Elfasol method is associated with the imaging of offset metal printing plates and is more of a laboratory process or a print room process rather than being suitable for operation in the administrative office.

Many of the steps in these processes are the same, some of the steps differ only slightly and only in a few cases are they radically different.

Before studying the sequence of events that make up the process it is important to understand something of the characteristics of the individual ingredients that go to make up the original process and the alternative ingredients that are used in the processes that have been derived from the original invention.

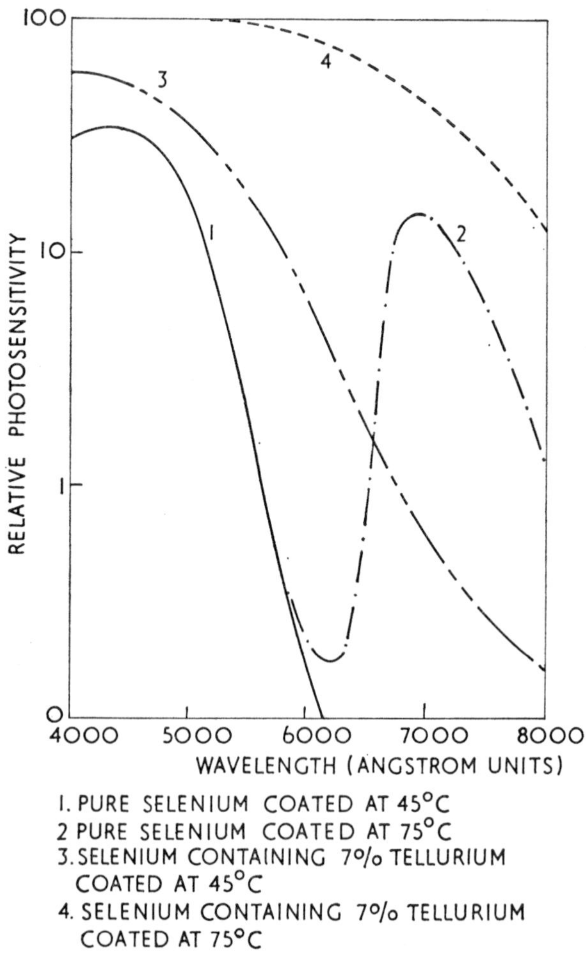

Fig. 6. The spectral response effect of additives to, and changes in the manufacture of selenium based photoreceptors. The curves are plotted for a spectrum of equal energy. (From Keck, *Journal Opt. Soc. Am*, Vol. 42, p 224, 1952). The object of research into photoreceptors is to make curve 4 as horizontal as possible.

Additives for photo-receptors

In the first place, there must be a photo-receptor. This is a surface upon which the light and dark shades of the original fall in the form of words, figures, line diagrams, tonal and half-tone pictures. The three best known and most widely used surfaces are basically amorphous selenium, zinc oxide and the Elfasol compound which seems to have different names in each country. Whereas all three photo receptors in their basic form operate successfully as electrostatic image forming surfaces, they all work better with certain additives. The precise nature of the additives has cost a lot of research and development money to discover and test. It is not surprising therefore that their sponsors are shy about revealing what these additives are. However, a little is known.

Tellurium is one of the selenium additives and there are others which are not specified. Various dyes can be added to the basic zinc oxide coating for papers used in the Electrofax process. These additives include acridine orange, fluorescein, eosin, rose bengal and methylene blue.

The effect of these additives is to enhance photosensitivity and extend the spectral response of the photo-receptor to the various wave lengths of light that are reflected from the original that is to be printed. The Kalle Company of Germany have imposed a complete blackout on the science of their photo-receptor coatings so information on this subject cannot be given.

Research and development is continuous and no doubt other additives have been, and will be discovered to make these surfaces reprographically more useful and able to produce better quality prints at faster response speeds.

Figs. 6 and 7 summarize the effect of additives to the selenium and zinc oxide photo-receptor surfaces.

Photo-receptor carriers

A surface, by definition, implies the existence of a carrier of that surface. The photo-receptor surfaces discussed above have, by their nature, to be supported on something.

The zinc oxide compound with its additives is flexible and is usually coated on paper. The photo-receptor thus becomes an expendable item. Commercially the method has both advantages and disadvantages, according to the context of the application, as in fact have all processes.

The Elfasol compound for reasons associated with the methods of image development rather than the nature of the compound, is carried on a flexble metal offset printing plate and once developed is expendable. The selenium is used in layers only 40–80 microns thick, vacuum coated in highly sterile conditions on aluminium plates or drums of various dimensions. The plates are used in manually operated equipment. The drums are used in automated copier/duplicators. The selenium surface is reusable.

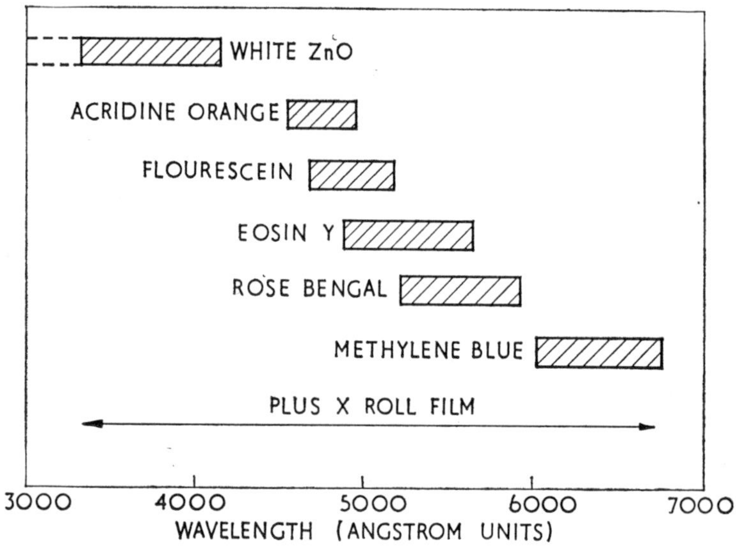

Fig. 7. The spectral response effect of additives to basic zinc oxide photoreceptors (from Young & Greig, *RCA Review*, Vol. 15, p 476, 1954).

These photo-receptors have in common the peculiar property of being able to change their electrical characteristics according to whether they are illuminated or not. Electrically they are classed as semi-conductors. They are also known as photo-conductors. More detailed consideration of these photo-receptors, their characteristics, how they work and so on is given in Chapter 4.

Producing the image

The next process ingredient is a device or mechanism for charging the photo-receptor with positive or negative static electricity and thus making it electrically light sensitive. This mechanism is common to all the three processes under consideration. It is called a charging bar by some, a scorotron or a corotron by others. It sprays static electricity at the photo receptor (see page 46).

When the charging bar has made the photo-receptor ready to receive the image, an apparatus is required to produce that image. This is, in effect, a lens system or camera. There is an arrangement to hold an illuminated original so that its image is projected on the charged photo-receptor. The image is produced by the charges in the non-image areas being dispersed by the action of light reflected from the equivalent areas of the original.

After exposure the photo-receptor is in a new electrical state. It is partially charged and partially discharged. The partially charged areas carry a voltage which varies according to the tonal quality of the original. Dark tones result in high voltages. Light tones produce low voltages as do actinic colours such as blue, while red tones result in high residual voltages. Those areas of the photo-receptor which correspond with the white areas of the original are, in theory, totally discharged.

Developing and fixing the image

The electrostatic image is invisible and it can be maintained only in darkness. Therefore it has to be developed in darkness (see page 35). The effect of development is to cause a fine pigmented thermoplastic resin powder called toner to stick electrostatically to the residually charged areas on the photo-receptor which represent the image of the original.

After development, the image can be transferred electrostatically from the photo-receptor to a non-proprietary media such as ordinary office stationery, and made permanent on this media; or the images can be made permanent on the original photo-receptor. In this case, the sensitized photo-receptor becomes an expendable item in the process.

The developed image is at this stage only holding to the chosen media by electrostatic attraction and therefore would smudge if touched. To be of practical use, it has to be made permanent. This can be done by heating the image so that the individual particles of toner which make up the image are melted together and into the upper layers of the media. Where liquid development is used (see page 74) evaporation of the liquid allows the toner particles to settle on the photo-receptor.

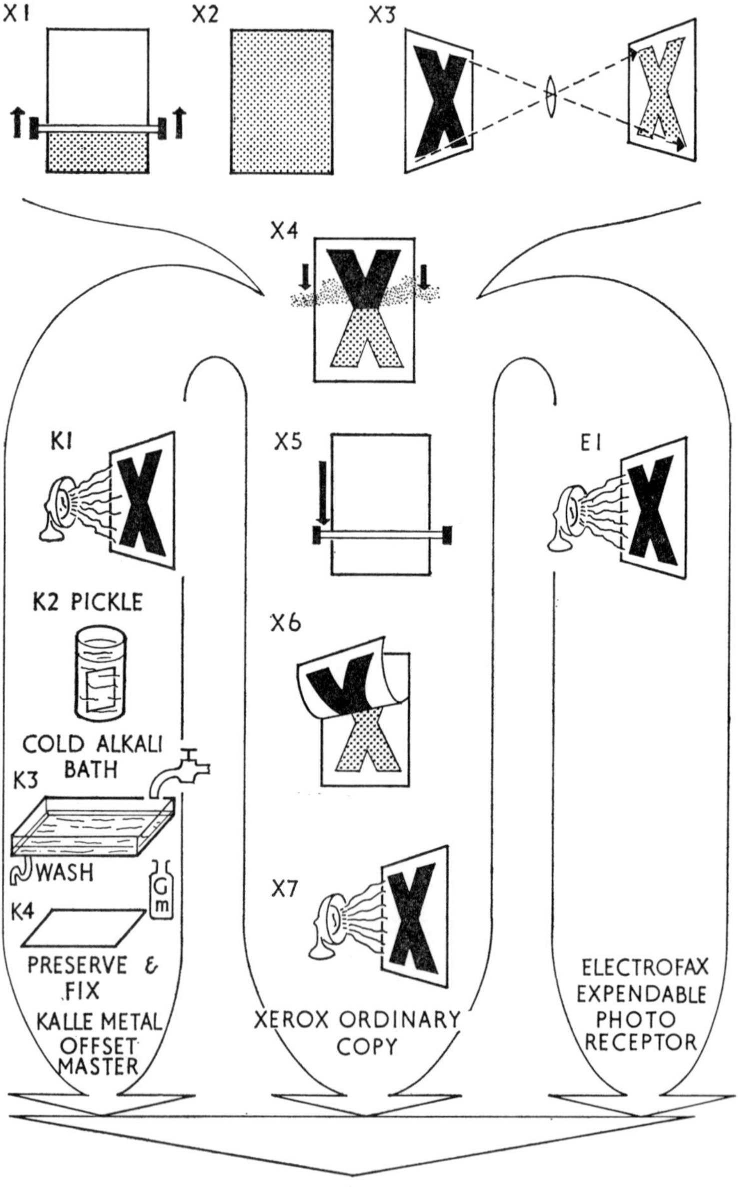

Fig. 8. The three xerographic methods of creating useful electrostatic images. The alternative ways of fixing the images on Electrofax papers are described later in the text.

Basic process and variations

The basic process with its derived variations is shown diagrammatically in Fig. 8. The first four steps are common to all the methods. The user then has the choice of one of the two specialized methods or the more general method that images ordinary paper and similar media.

The main stream method from which the other methods were evolved, has its steps numbered X1 to X7 in Fig. 8. The Electrofax derivation is labelled E1 and the Elfasol method's steps are labelled with a K serial in the diagram.

The first four diagrams, which are common to all methods, are described as follows:—

X1. The charging of the photo-receptor with static electricity by means of a charging bar. The polarity of the static electricity varies with the method and is of no consequence at this preliminary stage of explanation.

X2. The photo-receptor in its contrived electrical state after charging when it is ready to receive the image. The photo-receptor is held in this state in total darkness. In these conditions the surface of the photo-receptor maintains itself as an insulator and thus the static electricity is prevented from leaking away.

X3. The imaging stage of the process. Only the lens is shown; all the other camera paraphernalia such as lamps, iris and shutter, have been removed for the sake of clarity and simplicity. The object on the left is projected as a shadow on the charged photo-receptor. Naturally the static charges leak away in the non-image areas because in these areas the photo-receptor surface has been converted by light into a conductor of electricity. The static charges leak away to earth, so good earth connections are essential in all electrostatically operated powder printing machines.

It is at this stage that the variable or fixed reduction and/or enlargement ratios are applied in the machine design specification.

X4. The development stage where, by one means or another, the toner is brought into contact with the latent electrostatic image. The polarity of the toner is arranged to be of the reverse to the polarity of the residual image after exposure. The powder adheres to the electrostatic image areas thus making them visible and, depending upon the efficiency of the method, the materials and the machine maintenance, it does, or does not adhere to the photo-receptor in the equivalent blank areas of the original.

It is at this point that the derivatives of the original process emerge as separate xerographic methods of getting meaningful marks on to paperlike media.

E1. With the dry Electrofax method, there is only one more step necessary; the fusing step. Sufficient heat is applied to the developed image to melt the toner particles together and into the top layers of the base material of the photo-receptor. The copy can then be handled and used either as a copy or as a master for the offset litho duplicator, depending on the composition or the photo-receptor surface.

K1. The Elfasol method requires the development image to be fused *before* the other steps are taken which lead to production of useful output. This is because the developer consists of two powders, one of which has a lower melting point than the other. It is the lower melting point powder that is used to create the permanent and durable image. The other powder is there for technical reasons (see page 74).

K2. The fused master with its organic photo-receptor surface has to be pickled in an alkali bath, which contains volatile substances and should not be heated beyond normal room temperatures. The reasons for the alkali bath arise from the object of the method. The Elfasol method was evolved specifically to produce metal offset masters of good quality by using electrostatic charges as an alternative to the chemico-photographic methods also sponsored by the Kalle Company. The Elfasol master therefore is coated with the patented organic photo-receptor on which the tough offset litho image is to be built up. Once developed, the unused organic coating remains in the blank non-image areas and would interfere with the offset litho printing process. The pickling step is intended to remove the organic coating from the master in the non-image areas.

K3. The pickling liquid and the remnants of the powder that was not fused, have to be washed off; the washing stage is carried out in a normal sink with cold water supply and waste disposal. Paper-based offset litho masters cannot be produced directly by the Elfasol process because the double wetting would damage the fabric of the carrier of the photo-receptor surface.

K4. The carrier of the photo-receptor surface is an aluminium offset litho printing plate. Before processing, the surface is protected from contact with the atmosphere by the organic photo-receptor coating, but this is removed by steps K2 and K3 above. The bare aluminium surface therefore, has to be preserved and fixed by the application of a thin coating of a solution of gum arabic and phosphoric acid. This is normal procedure in offset litho operations when a newly created master is not going to be used at once.

Having disposed of the two methods derived from the original invention, it is possible to return to the main stream process at diagram X4, from which the other two methods branched off.

With the Xerox method, a reusable photo-receptor is used in the form of a plate in the case of manually operated equipment, or in the form of a drum in the case of the automated machines. At stage X4, the latent electrostatic image on the drum is powdered using the cascade development method (see page 72).

X5. As the powdered image emerges in the case of the automated machine, at the correct moment and in synchronism with the rotation of the drum, a sheet of normal office stationery is fed in so that it comes into contact with the emerging powdered image. Another charging bar applies static electricity to the paper of a polarity opposite to that of the powder and so transfers most of the powdered image from the drum to the plain unsensitized paper.

X6. As the drum continues its rotation, the paper, now with its powder image, is peeled away from the drum by methods which vary according to the design of the machine. The drum continues its revolutions passing a drum cleaning station where the surface is prepared for another cycle involving charging, exposure, development, transfer, clean and a repetition of these functions.

X7. The paper with its powdered image passes under a fusing device where heat is applied so that the individual toner particles fuse together and into the upper layers of the paper. The paper is then carried to its delivery point.

Manually operated equipment works in a similar fashion. The plate is charged in a special compartment of the processor and at the end of the charging process it is shielded while being manually transferred to the camera. The light shield is removed and an exposure made. The plate is shielded again and transferred to a development tray which contains the developer powder. It is clipped to the tray, the light shield is removed and the tray with the plate attached to it is rocked to and fro so that the developer rolls backwards and forwards across the plate. This allows the toner to adhere to the image areas on the plate. The plate is removed by hand from the developing tray. The unsensitized media which is to be imaged is positioned on the plate and both are inserted into the charging compartment of the processor where a positive charge is spread all over the paper or other media. The media is peeled off the plate, and transferred to the heat or vapour fusing compartments. This makes the image permanent. The plate has the residual image dusted off and is returned to the storage compartment for reuse. On being taken out of the fusing compartment, the copy is ready for immediate use.

There are two extra stages when using the Xerox method to print xerographically on normal paper or other media in everyday office use when compared with the Electrofax process. This is sometimes put forward as a criticism of the Xerox process. With a process involving a rotating drum, however, it really does not matter how many stations of operations there are around the periphery of the drum. The speed of output depends upon the speed of the rotation of the drum and not upon the number of operations taking place simultaneously around the drum. Automated machines using the Xerox method now operate at speeds of 3,600 impressions per hour.

Electrographic processes

The two processes that go to make up the electrographic group of powder printing processes are called the Videograph process and the Videojet printing process. The words Videograph and Videojet are trademarks owned by The A. B. Dick Company.

Videograph. The key element of the Videograph process is a specially designed cathode-ray tube in which an electron gun is sealed and so mounted as to direct an electron beam toward a flared rectangularly-sectioned array or matrix of spaced, small diameter wires fused into the opposite end of the tube. These wires extend across the major dimension of the tube, their inner ends being directed towards the electron gun. Their outer ends are exposed to the atmosphere so that the wires form a series of spaced and insulated conductors extending from the inside to the outside of the tube.

The matrix is formed by a series of rows of stainless steel wires of 0·001 in. diam. spaced on 0·004 in. centres. The vertical height of

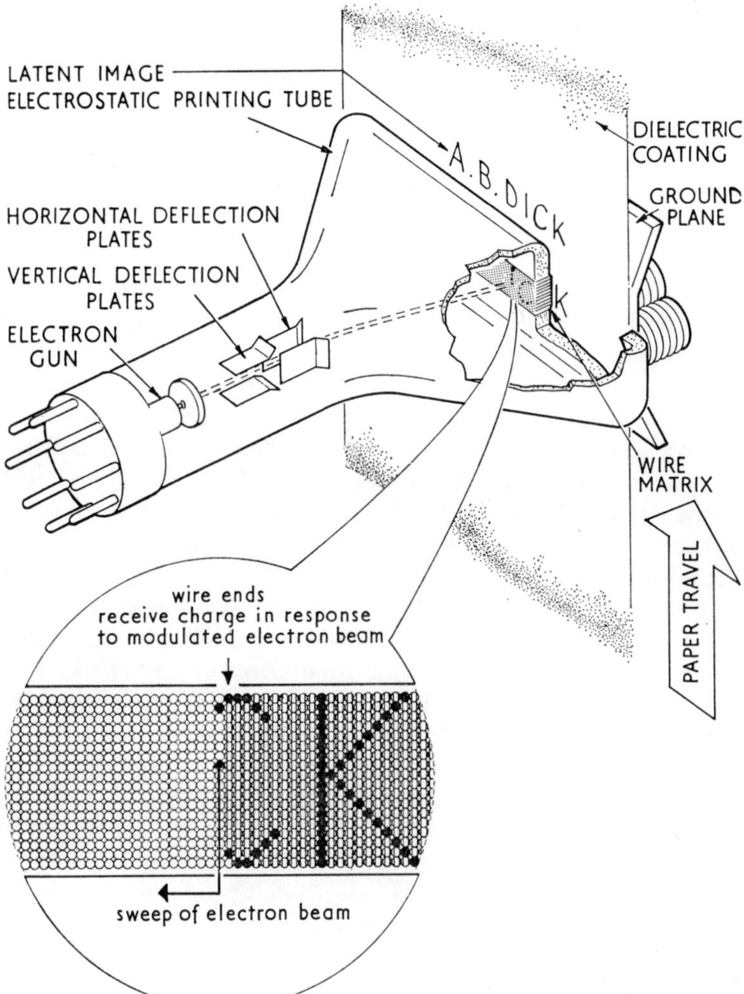

Fig. 9. The A.B. Dick Videograph method of directly imaging a moving web of Electrofax or similarly coated paper with electrostatic charges.

the matrix is such as to print the maximum height of the characters likely to be required.

Fig. 9 shows how the latent electrostatic image is created on the web of sensitized paper that passes through the unit at 180 ft. per minute.

As the paper passes over the matrix, it is pressed to an earthed plate in order to improve image definition and reduce background.

The image of the original has to be transformed into a series of "X" and "Y" voltages which is applied to the deflector plates so that the electron gun pulses of electric charges are directed to the appropriate wire or wires in the matrix. They are then deposited on the proprietary brand of specially coated or sentitized paper.

The electrostatic image imposed on the special paper is developed and fixed in a way similar to that described for the Electrofax method (see Fig. 10).

Videojet. The Videojet printing process works on radically different and new principles to cope with very fast speeds of character recognition from electronic sources.

When a pressurized fluid is ejected through an orifice it produces a stream of irregularly-spaced, non-uniform droplets. However, if the orifice is energized by ultrasonic vibration by means of a coil or a piezoelectric crystal, the droplet stream becomes uniform with respect to the size of the droplets and their spacing. Their rate is a direct function of the energising frequency. Thus, if the energising frequency is 75 Hz, the stream comprises 75,000 droplets per second.

If a video signal is applied at the orifice, each droplet accepts an electrical charge equal to the instantaneous value of the video signal at the point where it is formed in front of the orifice. The Videojet printing technique utilizes this principle for directing a controlled stream of ink droplets against non-proprietary paper or other printing media to produce high-quality, electronically controlled, non-impact printing. For alpha-numeric printing, an electronic character generator in the video circuit controls the ink stream to form characters based on a 9 × 11 dot-pattern. A video signal representing the white or blank spaces between characters deflects the ink particles away from the media to be imaged into a waste container.

The ink used in the Videojet process is similar to ordinary fountain pen ink and may be provided in a variety of colours. A special ink can be provided to prepare direct offset masters which can be readily duplicated in conventional offset duplicators.

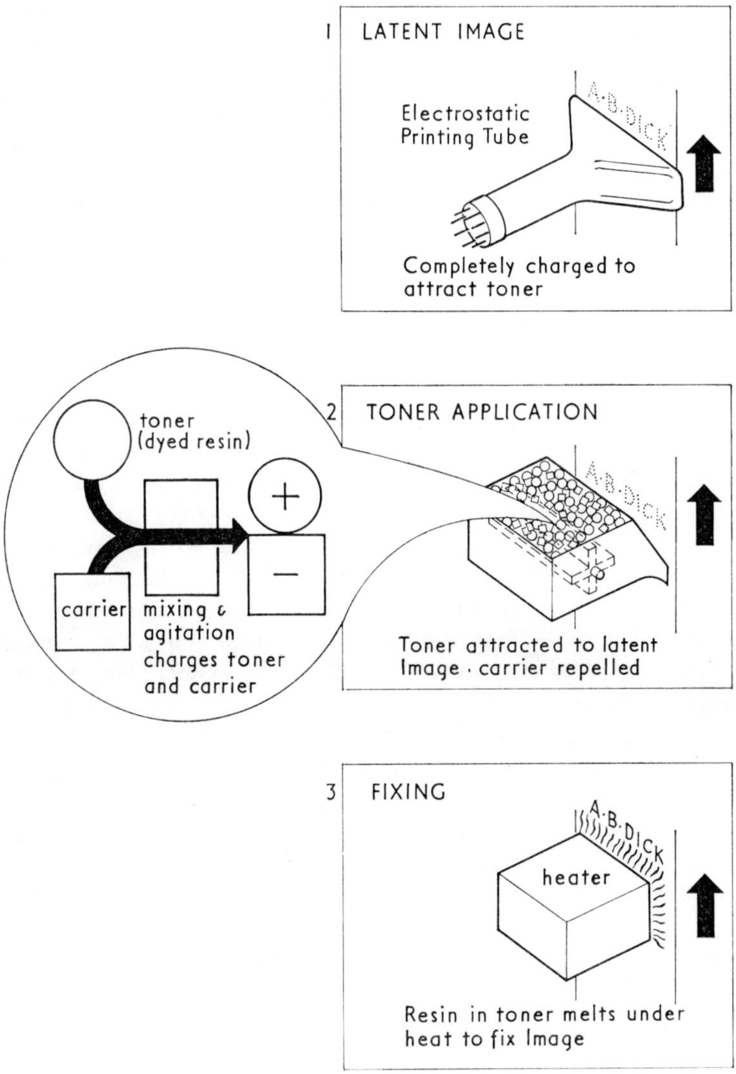

Fig. 10. The A.B. Dick Videograph method of developing and fixing the electrostatic image created electrographically.

The 960 Series printer is a console unit of the pedestal type. The ink jet as described above is mounted in the head assembly of the unit for reciprocation across a vertically-moved paper transport system by means of a friction-drive belt. Printing occurs as the ink jet is moved from left to right across the paper at a synchronous rate of four milliseconds per character, or 250 characters per second. The jet assembly is automatically returned to the left-hand starting position at any point in its travel in response to the receipt of a carrier return signal. Return travel is three times faster than the writing speed, or at 75 in. per second.

The printer accommodates edge-punched zig-zag folded paper or card stock from 3 in. to 14¾ in. in width, and from 15 to 125 lb. ream weight. Paper advance is on a line-by-line basis at a speed of 16 milliseconds per line in response to the receipt of line advance codes.

The electronic logic of the printer is modularised and employs solid state circuits throughout. The upper assembly of the printer includes the mechanical printhead, ink supply system, and the electronic components associated with the printhead. The pedestal includes the solid state character generator board providing 64 character repertoire, the power supplies, and an interface board adapted to directly interface with (a) Bell 201 data sets or equivalent, (b) Bell 202 data sets or equivalent, or (c) an associated buffer.

By reason of the above modularisation, the printer can be supplied with an appropriate interface board for any system configuration; without the interface board and character generator boards for use as a slave printer; or without interface boards where series parallel conversion, error detection, etc., are provided in a common system control unit.

CHAPTER IV

HOW THE PHOTO-RECEPTOR WORKS

It is well known that certain elements and compounds have the characteristic of conducting a flow of electrons, and that certain other elements and compounds act as a barrier to this flow of electrons which we call electricity. The barrier type elements and compounds are known as insulators and the carriers of electrons are called conductors. A third group of elements or compounds—called semi-conductors—can act in either capacity, given the necessary inducement. When the inducement to change its electrical behaviour is the introduction of light energy, the semi-conductor elements and compounds are also known as photo-conductors. The phenomenon is known as photo-conductivity, first noted by May in 1873 and demonstrated by W. H. Allwachs in 1888.

But for the existence of elements and compounds which can react to light and have this ability to change from a conductor to a non-conductor of electricity, there would be no electrostatic printing processes. There would, however, still be the electrographic methods of printing.

Electrical qualities

Photo-receptors can be of zinc-oxide, selenium, sulphur selinide, cadmium sulphide and various other types, but they all work in a similar fashion. Basically three things happen to a photo-receptor:

1. From a neutral state of rest, it receives and holds an electrical charge of negative or positive polarity.
2. It selectively releases the electric charge according to the instructions it receives from the pattern of light and shade that falls on it.
3. It has to recover its original receptivity to a new electric charge for it to repeat the stages mentioned above if it is required to be re-used for repeated copies and duplicating operations.

It follows that the photo-receptor has to exist, at least momentarily, in two electrical states at one time. It has to be partially in the state of an electrical insulator and partially in the state of an electrical conductor. It must be able to conduct electricity *one way* and in one direction, and not leak electricity from image area to non-image area laterally.

In order to understand how it does this, it is necessary to have at least an elementary understanding of the structure of the atom.

Atoms and their electrons

Essentially, all atoms, irrespective of the nature and characteristics of the element or compound in which they exist, have certain similarities. They all have a nucleus and they all have a cloud of electrons orbiting around the nucleus. Electrons are minute particles, in that they have a mass that has been measured, and an electrical charge which has been labelled negative. The mass and the negative charge of these electrons are almost inconceivably minute. The fact that a figure for the mass has been established proves that the electron is corpuscular.

Within the nucleus around which the electrons are orbiting very fast, there are two particles; the proton and the neutron. The proton is again corpuscular for its mass has been established as being 1840 times the mass of the electron. The proton also has an electrical charge which has been labelled positive.

The neutron which is also contained within the nucleus is another particle. Its weight is about the same as a proton but it has no electrical charge.

These are the only units of matter which need concern us in our understanding of how a photo-receptor works. In fact we need only concern ourselves with the electrons and leave all the other fascinating aspects of the anatomy of the nucleus to the academic physicists. For our purpose, the proton and the neutrons remain immovably fixed within the nucleus.

In its natural, inert pre-charge state, all the atoms of the photo-receptor are electrically neutral. That is to say, each atom has an equal number of electrons, protons and neutrons. In the diagrams which follow, an atom is drawn as having three of each particle or charge. This is done in order not to clutter up the diagram. In most atoms and certainly in the case of the atoms that make up the photo-receptor, elements or compounds each contain very many more electrons, protons and neutrons than just the three shown.

Charging the photo-receptor

The picture of just one atom of the photo-receptor is represented in Fig. 11 in simplified form, in a neutral state. The negative and the positive charged particles balance out and the electrically neutral charged particles do not affect the issue. The atoms of the photo-receptor are electrically inert so the photo-receptor as a whole is inert.

The charging stage is shown in Fig. 13. One electron has been added which electrically unbalances the system and so the atom

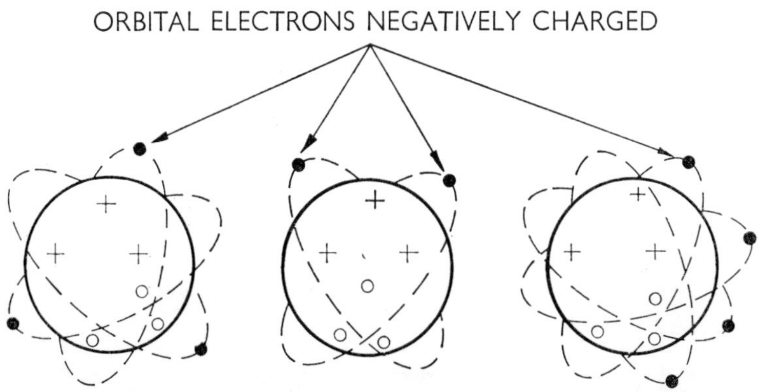

Fig. 11 A neutral atom

Fig. 12. A positively charged atom

Fig. 13. A negatively charged atom

becomes negatively charged. Similarly in Fig. 12 an electron has been taken away from a neutral atom with the result that the atom becomes predominantly positively charged.

Inspection of these diagrams shows that the number of protons and neutrons does not change but the number of electrons does change; so does the electrical balance of each atom and therefore the photo-receptor as a whole.

It follows that to charge a photo-receptor, a device is needed that will emit electrons or attract electrons. Fortunately such a device is fairly simple in concept. It is called a charging bar in lay language or more technically, a corona discharge, or coratron or scoratron.

In Fig. 14, the inner wires on the narrow arc are the ones that either emit or absorb electrons. The other wires on the larger arc are field stabilizers and controls. In some charging bars a metal screen takes the place of the wires attached to the larger arc. For our purpose, we can ignore these finer points. It is easier to understand what is happening if we consider an end on view of a charging bar (see Fig. 15).

What happens when the charging bar is switched on depends on the fact that when a wire with a very high degree of surface curvature is loaded with a very high direct current voltage, the wire glows white. It has been called the St. Elmo's fire effect. In fact, the wire is not

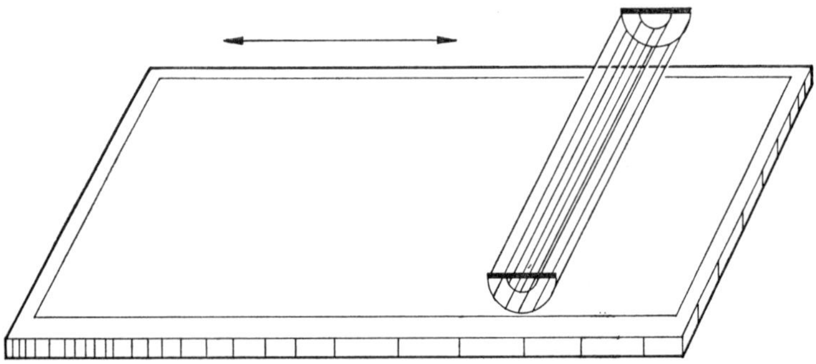

Fig. 14. A charging bar (or scorotron) positioned close to a photoreceptor which could move either way. Alternatively the photoreceptor could remain fixed while the charging bar moves. The photoreceptor could also be a segment of a rotating drum.

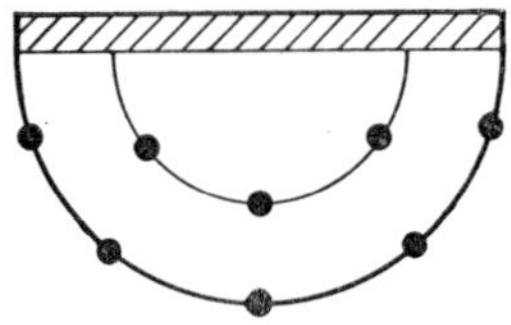

Fig. 15. An end-on view of a charging bar or scorotron.

getting hot; the air around the wire is becoming ionised. The air atoms become electronically unbalanced, loaded with excess electrons or deprived of electrons according to the polarity of the high voltage used. The high degree of curvature of the wire is obtained by using very fine wires.

In practice, the coratron wires seldom need to be so fine, or to carry voltages so high that they glow. The precise thickness of the wires and voltage loading are commercial secrets which need not concern us.

The process of charging a photo-receptor surface in the dark is illustrated in Fig. 16. On the left, the charging bar is assumed to have been positively charged. Air electrons are being absorbed and the electron deficient atoms are stealing electrons from the adjacent atoms and so on until the atoms are in contact with the photo-receptor. Only two of the millions of atoms are shown. The last air atom in the chain grabs an electron from the photo-receptor. Being in the dark, the surface is non-conductive and the photo-receptor electron-deficient atoms cannot replenish their orbits with electrons. They therefore remain in a positively charged state.

Should it be necessary to charge the photo-receptor negatively, the reverse happens. The charging bar adds electrons to air atoms which hand them on ultimately to the photo-receptor atoms which, being in a non-conductive state in the dark, cannot hand on the excess electrons and so become negatively charged.

Discharging the photo-receptor—theory

Having understood how a photo-receptor can become negatively or positively charged, we can now investigate how it becomes selectively discharged, or for that matter, how it loses its charge when light falls on an electrostatically charged surface.

It is common ground that the electrons of the atom are circulating around the nucleus in a sort of haze or cloud. Just as with our atmosphere the rain clouds are sometimes high up and sometimes at ground level, so this cloud of electrons is positioned relatively close to, or further away from the nucleus.

In certain elements or compounds, the natural position of this cloud is at one level while with others the area of orbit is at another level. For the sake of convenient understanding, let it be assumed that in the case of electrical conductors, the cloud of electrons is orbiting high up or at a significant distance from the nucleus, whereas with insulators the cloud is significantly closer to the nucleus.

Unlike weather clouds, the electron clouds remain fixed in their general orbiting area in the case of the two classes of atoms just mentioned.

In the case of the photo-conductors however, the very nature of the element or compound permits the cloud to change its orbiting area given the necessary inducement. It is as if lightning causes the clouds to lift.

When light strikes the photo-conducting atom, the electrons

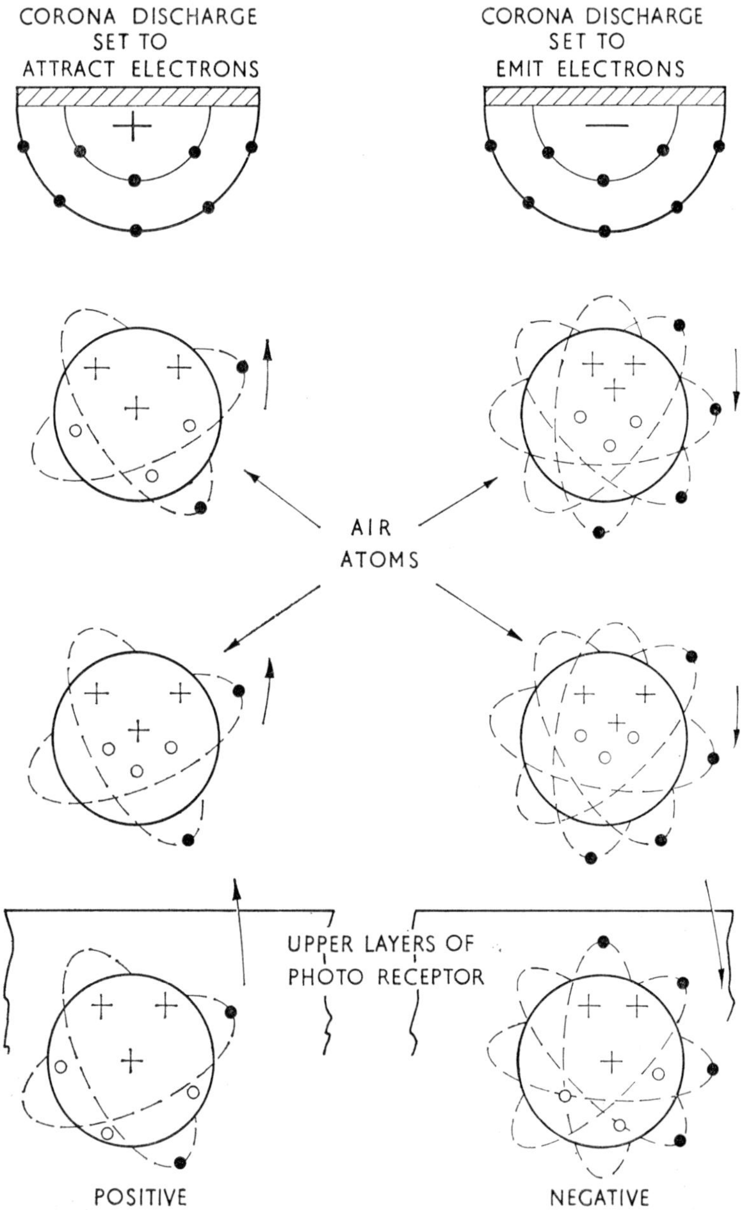

Fig. 16. The photoreceptor shown as receiving a positive or negative charge by having electrons removed or added to the surface atomic structure.

change their station and rise above the level they would be if the atom was in the dark. Alternatively, it can be envisaged that it is light that excites the electrons so that they rise to new heights and in darkness they fall back into a less energetic orbit. Conductors cannot do this, neither can insulators, otherwise these substances would be known as photo-conductors.

Without getting too involved in the strict science of the system, let it be supposed that in their high orbit the cloud of electrons can be moved on under the influence of the "wind", or pressure of an

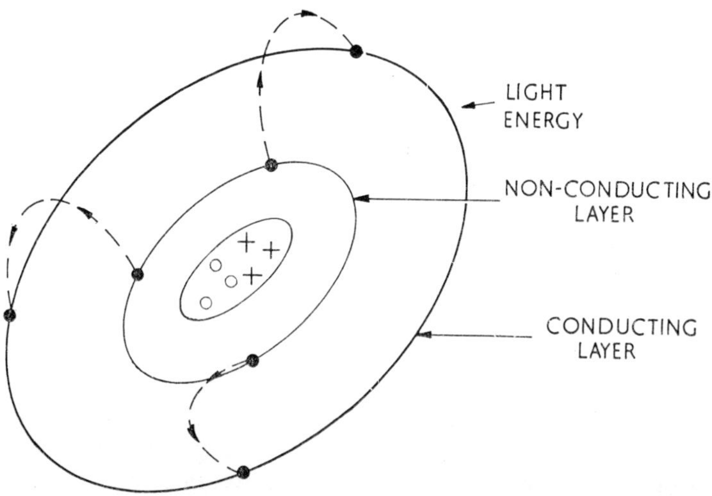

Fig. 17. Light energy is seen to be entering the atomic structure of a photo conductor and causing some of the electrons to be shifted into an orbit from which they can be moved on to adjacent atoms under the influence of an electromotive force.

electromotive force or voltage difference. Conversely, let it be supposed that when electrons are in a low orbit near the nucleus they are not susceptible to the "wind" of an electromotive force. The factor that causes the change of orbit is the introduction of light energy. This energy induces the electron cloud to rise to a point where it becomes less attracted to the nucleus and so can be moved on to the next nucleus and so on. This graphically describes in a non-scientific way what happens when light strikes an atom of the photo-conductor and is illustrated at Fig. 17.

It may seem strange that the charge on a photo-receptor is prevented from dispersing before exposure, yet is allowed later to flow one way and not the other. In one case there is an electron deficiency (positive charge) and in the other case there is an electron excess (negative charge). This unbalanced state of affairs must not be allowed to be neutralised by the egress of random electrons on the one hand or the escape of electrons on the other. The one-way control vitally necessary for the electrostatic process to work is accomplished by the use of a special binder layer compound positioned between the photo-conductor and the substrate which carries the photo-receptor surface. In effect, the binder is an insulator to the flow of electricity in one direction, i.e. negative to positive, and not in the other direction, i.e. positive to negative. The nature of the charge polarity determines the nature of the binder compound.

A simplified explanation as to why charges do not escape from a charged area laterally to an adjacent non-charged area, arises from the fact that the orbital electrons in the image area have not been lifted by the action of light to their conductive layer. In the illuminated areas the action of light is to provide carrier energy to cause the electrons in these areas to disperse and as there are no electrons in the conductive belt of the atoms in the image area available to replace them the charges in the image area are not dispersed.

So much for the theory in a very simplified form. Simplification has inevitably introduced some technical inaccuracies but in this general work no attempt has been made to be strictly and scientifically accurate. The recommended reading adequately covers the detailed and scientific aspects of the process more accurately for those who want it.

Photo-receptor characteristics

In practice, the efficiency with which the theory works depends on many factors, some of which are:

1. Residual impurities which cannot be extracted from the photo-receptor in manufacture.
2. Choice of barrier compounds.
3. Method of manufacture.
4. Tolerances within which it is possible to effect commercial manufacture of the photo-receptors.
5. Methods of coating the substrate.
6. Conditions under which coating takes place.

How the manufacturers succeed in producing photo-receptors which work so well is, of course, their business and these details are not for publication.

Commercially-sponsored research and development into photo-receptors is continuous. A new type being developed (see page 133) is of multi-layer formation as distinct from the dual layer type so far mentioned.

The plate on which the electrostatic image is formed consists of a conductive base, a photo-receptor and an insulated surface. The formation of the image arises from the following sequence:

1. The top insulated (Mylar) surface gets charged and this induces a reverse charge on top of the photo-receptor.
2. The image is projected simultaneously with an AC corona discharge, which induces a charge pattern equivalent to the image on the top insulating surface.
3. The pattern is held until the image is developed and transferred to the unsensitized media required. The image when it requires to be erased can be dispersed by irradiation with an infra-red radiation source.

Zinc oxide photo-receptor

Zinc oxide compounds are flexible and can be coated on to a flexible carrier such as a paper base. The composition on the coating varies according to the purpose for which it is required. The cost varies, too, with the degree of sophistication and versatility that the manufacturers of equipment demand. Some zinc oxide papers can be used without further treatment as short run masters for offset litho duplication. Some can be written on with ink. Some can be marked with such unlikely objects as a coin and so on.

The resolution and sharpness of images varies between zinc oxide papers of different types and, subject to the additives, their colour response and image quality also varies. Some can produce colours in many shades of grey, others can produce only a few shades of grey and images tend then towards soot-and-whitewash rendition of half-tone images.

They all seem to cope with solid areas very well although with some the edges of these solid areas and heavy typescript may not be so sharp as with the more sophisticated papers. Development methods can affect image quality and it is important to use the right and recommended papers for the machine.

There is not much point in discussing photographic response speeds in view of the different characteristics of the paper in use. These are, in any case, only of academic interest. It is sufficient to note that the machines using these papers are set for the correct exposure for the recommended paper. The machine suppliers provide instructions for variations of exposures over narrow limits for different types of original if the machine design incorporates this feature.

Most, if not all, zinc oxide papers need a negative charge to make them light sensitive. Again this is a machine setting, so need not concern the practical reprographer.

The robustness of the zinc oxide photo-receptor is a function of the paper carrier but if the surface is fingered, invisible body-grease marks tend to attract developer or toner in unwanted areas. The paper must not be creased or damaged. Shelf-life is pretty long, but as most suppliers provide the paper in cut sheet or roll form for certain machines more or less as required, users need not and should not carry excessively large stocks for local needs.

The re-usability characteristics of the photo-receptor do not arise because the photo-receptor is the copy or the offset master.

Zinc-oxide receptors do not reproduce orange or red very well but otherwise the colour response is good.

Elfasol photo-receptor

This is certainly robust because the carrier of the coating is a metal offset printing plate but the surface could be scratched by careless use. The Elfasol plate can be reused if it is completely erased before it is fused. Such a need might arise owing to a wrong exposure. However, the Elfasol plate needs resting for some hours before reuse so that it can recover its full photo-response characteristics at normal temperatures. If the plate is heated to (say) 180°C it can be reused within 15 seconds or so.

Although this is an electrostatic photo-receptor, it is not usually associated with short run copying. It is, however, a medium for long run duplicating methods only when used in conjunction with an offset litho duplicator or printer. Print runs well in excess of 100,000 are claimed by the manufacturers. Before being run on the offset litho machine the surface has to be decoated in the non-image areas (see Fig. 8).

One type of developer is required for fine lines and another for heavy solids. When properly processed the results are very good.

The Elfasol plate's colour response is good, except when yellow appears on the original. Its photo-response speed is academic because only the correct exposure (about 30 seconds at *f*9 for 1:1 reproduction with a lot of light) will produce optimum results. As with the zinc oxide and indeed with others, the photo-receptor does not respond to light as does a photographic emulsion. It is the volume of high energy light that has to be more or less constant and this is worked out by the manufacturers of the equipment using their recommended photo-receptor.

The charge used to make the Elfasol photo-receptor light sensitive is negative, but this is of only academic interest. The polarity of the charge depends on the nature of the photo-receptor and is chosen to give the best results.

Furnium and Xerox photo-receptor

These are dealt with together because for all practical purposes they behave similarly and have the same merits and characteristics. Furnium is the photo-receptor compound used in the Statikon copier.

They are both reusable surfaces. Once created on the photo-receptor the toner image is transferred to ordinary uncoated paper. The residual image is then wiped off and the photo-receptor is re-sensitized for a repeat operation. This means there is a cycle of operations and implies that the photo-receptor is coated on a drum or cylinder which rotates.

The surface can be scratched by misuse but otherwise is durable. Depending on the nature of the photo-receptor and its maintenance, its life varies from 20,000–100,000 copies.

For the reasons given in the previous case, its photo-response speed is immaterial. The machines' design sees to it that it is used under the conditions best suited to the photo-receptor used.

These types of photo-receptor tend to reproduce blue as a rather pale grey unless something is done to counteract this weakness. In some machines, the original is illuminated with green light, in other cases additives are incorporated in the photo-receptor surface. Other colours are reproduced in four or five shades of grey. This colour response suggests that two or three colour printing should be possible but the possibility does not arise simply because at the time of going to press toner powders are being produced only in black. Two or three colour printing no doubt will come in due course, when the market's demand is strong and intensive enough to pay for the redesign and development costs of the necessary hardware. The market's demand must also develop so that the mass production of coloured toners is economical. The technological problems have been, or are about to be, solved.

The most interesting difference between the zinc-oxide photo-receptors and those at present being considered, is concerned with the "edge effect" or the "soft centre effect". This is a term freely bandied about to describe the image of a solid area of black on the original that is produced by these types of photo-receptor. A solid area of black measuring a quarter inch square or more is not reproduced as solid black. It has a sharp edge and away from the edge the toner shades off to quite a light grey. With great *elan*, the suppliers of zinc oxide papers claim that this does not happen with their method.

All electrostatic machines, however, have difficulties when originals combine fine lines with heavy solids. Heavy solids can be filled in at the expense of thickening up the fine lines; while sharp letters and figures can be achieved at the expense of having thin solid areas. The processes under discussion are intended for business, commercial and administrative operational purposes, not for prestige printing purposes. The technology for filling solid areas with these types of photo-receptor is known and, if it is important, machines incorporating this feature will appear, as they have done in the USA.

Cadmium sulphide laminated photo-receptor

Under this heading, only the claims made by the sponsors—Canon Camera Co., Inc., of Tokyo, Japan can be recorded (see page 133). No machine has yet appeared on the European market.

It is a reusable photo-receptor, so its image can be transferred, as with the Xerox process, on to plain paper, transparent film and metal plates. The implication therefore is that rotary machines will ultimately be designed and marketed provided that no patents are infringed. It produces a high contrast image and is said to give no edge effect. It can stand 30,000 consecutive transfers. It is claimed to have ten times higher sensitivity than any process on the market in 1968.

It is difficult to see what this new photo-receptor does that established types do not. It can hardly be expected to produce better print quality than is at present produced. When machines using it appear, their success will depend upon whether they can produce excellent copy quality more quickly at a significantly lower cost per copy without sacrificing convenience, reliability and ease of operation.

CHAPTER V

IMAGING THE PHOTO-RECEPTOR

The optical arrangements in most electrostatic copiers and duplicators are similar and are adaptations of the systems used in cameras. All machines have a place to locate and illuminate the original, a position to accommodate the photo-receptor and a light source. Most machines have a lens, but there are some lens-less models (see page 64). Some of the machines which use a lens system incorporate a mechanism for changing the position of the lens in relation to the object (original) and image planes. Thus, enlargements and reductions become possible. In practice, only a few machines have this facility built into the design, and then can only give effect to reductions. The manually operated and semiautomated machines incorporate both enlargement and reduction facilities.

Function of the lens

The basic principle is illustrated in Fig. 18. The lens is shorn of all peripheral paraphernalia such as aperture control and shutter mechanism for the sake of simplicity. The lamps are only indicative of the light source. In many machines they are strip lamps. The light energy is shown as striking the original, passing through the lens and so producing an inverted shadow of the original image on the photo-receptor. The charges are seen to be leaking away from the non-image (white) areas to earth.

The charge polarity shown is positive. If the original charge is negative, a negative charge remains in the dark or image areas, and the negative charges in the white areas are dispersed.

It is vitally important to incorporate a lens in those machines that image normal paper because only light can be allowed to contact the photo-receptor in its charged state. If anything else is brought into contact with a charged photo-receptor, it unbalances the charge density and can partially disperse it. This also applies to the lens-less machines which use Electrofax paper.

The nature of the lens used depends on many factors concerned with the design and dimensions of the machine, the nature of the lighting, characteristics of the photo-receptor and so on. The lens in automated machines is usually of the wide angle fixed focus, fixed aperture type with very limited depth of field. The user cannot adjust it.

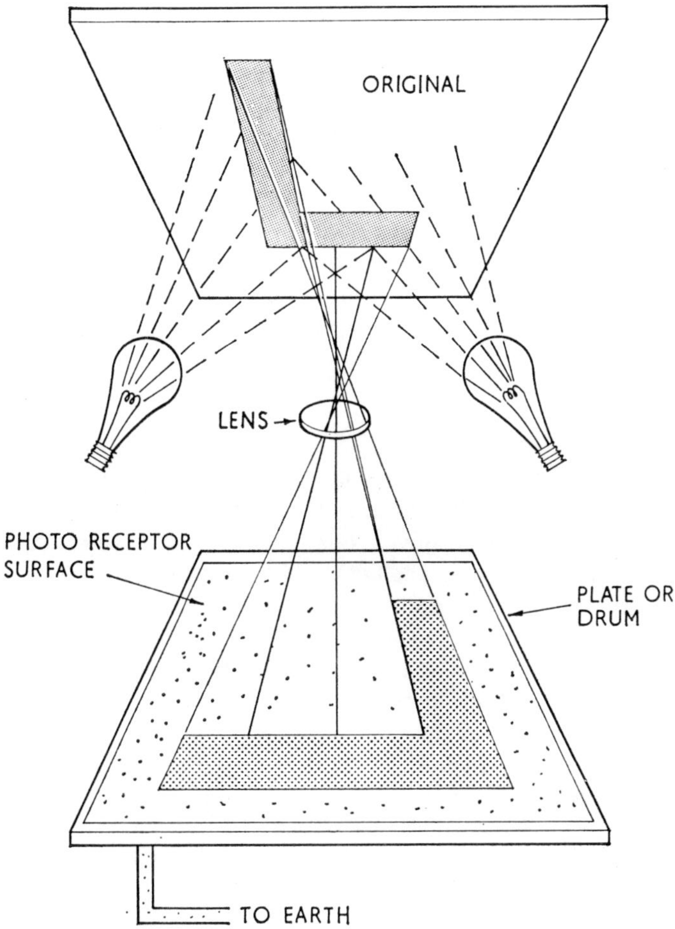

Fig. 18. The latent electrostatic image being built up in the photoreceptor.

Additional function of the lens

In addition to the normal function of a lens assembly, the lens in electrostatic equipment collects and distributes light energy (photons) *volumetrically.*

It is vitally important that the amount of light energy reflected from, or projected through the non-image or white areas of the original is enough to convert the equivalent areas of the photo-receptor into an electrical conductor. Very approximately one photon causes one electron to migrate. Too much light may cause the image to be weak, especially if the original image is not absolutely opaque, and may induce a residual charge in the white area to be built up as a result of the excess energy reaching the photo-receptor. Thus, when the photo-receptor is developed, this residual charge may attract toner and cause background. In this case the image to background contrast is poor. Too little light only partially discharges the static electricity in the white areas and again causes background at the development stage. Too little light may also cause excessive toner deposits on the image areas with the result that fine line resolution may be lost.

Control of light intensity

This compensation could be arranged by dimming or brightening the lamps but control by means of the exposure time is easier, specially as exposures are relatively long—30 seconds or so at *f*22 with a camera setting of 1:1. The light volume can thus be metered with the necessary degree of accuracy by the timer which is usually part of the camera equipment. The objective is to arrange for the amount of light falling on the original to be the amount that the photo-receptor needs to operate efficiently, irrespective of the lens setting used.

Another factor which alters the intensity of light falling on the photo-receptor is the reduction (or enlargement) ratios applied in the path of the light from the original to the photo-receptor itself. A reduction factor concentrates the light in the white areas, so exposures tend to be shorter for any given aperture setting. The table below illustrates the association of these factors and refers to a good contrast original when the Rank Xerox 1385 master maker is used.

The exposure times vary with the method of development, the formulation and condition of both the photo-receptor and the

RELATIONSHIPS BETWEEN REPRODUCTION RATIO AND EXPOSURE

Aperture setting (*f*-number)	Exposure setting (seconds) at reproduction ratio: 1 : 2	1 : 1	2 : 1	1 : 2	1 : 1	2 : 1
	Line subjects			*Tone subjects*		
11	—	8	12	7	13	20
16	8	15	23	13	25	36
22	15	30	45	23	45	68
32	30	60	90	40	80	120

developer and the positioning of the illumination sources. Separate figures are given for line and tone subjects because different development methods are used. Tonal subjects need more intensive development with the result that background would tend to increase unless exposures are increased. The weakening of the image due to the longer exposures is compensated by the more intensive development method used. The exposure times are altered according to the quality of the original. There are numerous other anti-tonal and intensification development techniques, all of which need changes in exposure times. These are described in the product literature supplied to users.

With the manually operated method, the skill, experience and judgment of the operator can be brought to bear. With automated equipment however, simplicity of operation using press button techniques requires all the settings to be fixed. When manually operated, the various stages of the operation are isolated and operations can be interrupted or modified at any stage. Before fusing, for example, the operator is able to spot out blemishes reproduced from the original. This is not possible with automated equipment. Although very versatile and capable of imaging unsensitized acetate sheets, rubber stereos, printed circuit blanks as well as offset masters and plain paper, the manual process is slow. One offset master or three or four plain paper copies in three or four minutes is a fair rate of production for an experienced operator working with uncomplicated line originals which do not require much camera work.

Use of mirrors for right-reading images

When an image passes through a lens, it is inverted and laterally reversed. As the machines work positive to positive, requiring a right-reading image from a right-reading original, it is evident that

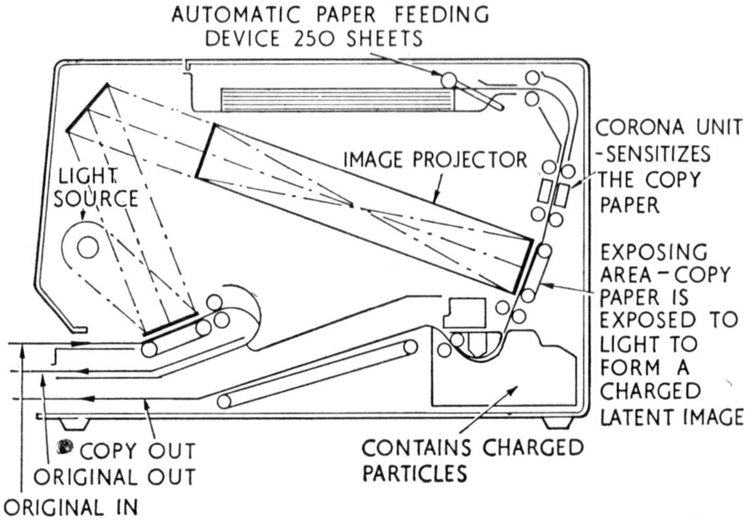

Fig. 19. Schematic diagram of SCM Model 33 Electrostatic Copier.

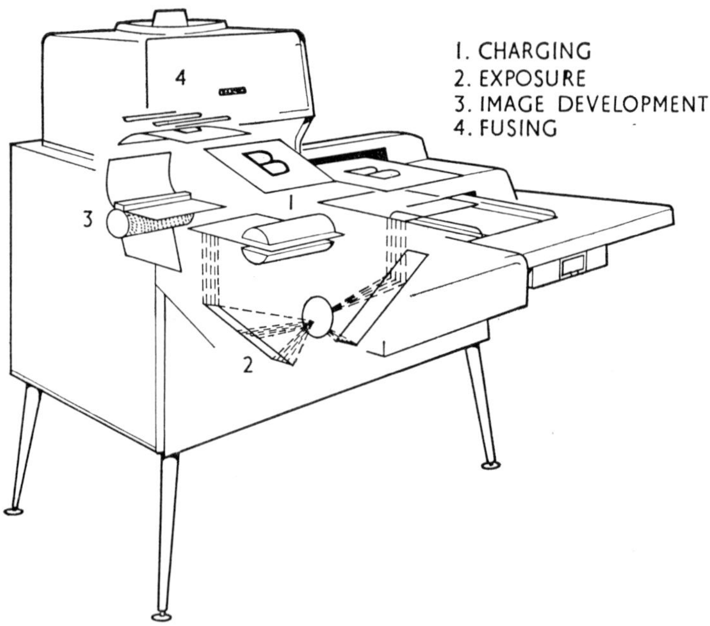

Fig. 20. Schematic diagram of the AM-Bruning 2000.

a mirror has to be used. Then again, to make these machines as compact as possible, the light rays have to bend round corners; mirrors are again used to do this. As one mirror would reverse the image again, two have to be used, so design complications increase. Quite obviously, none of these optical links can be allowed to intro-

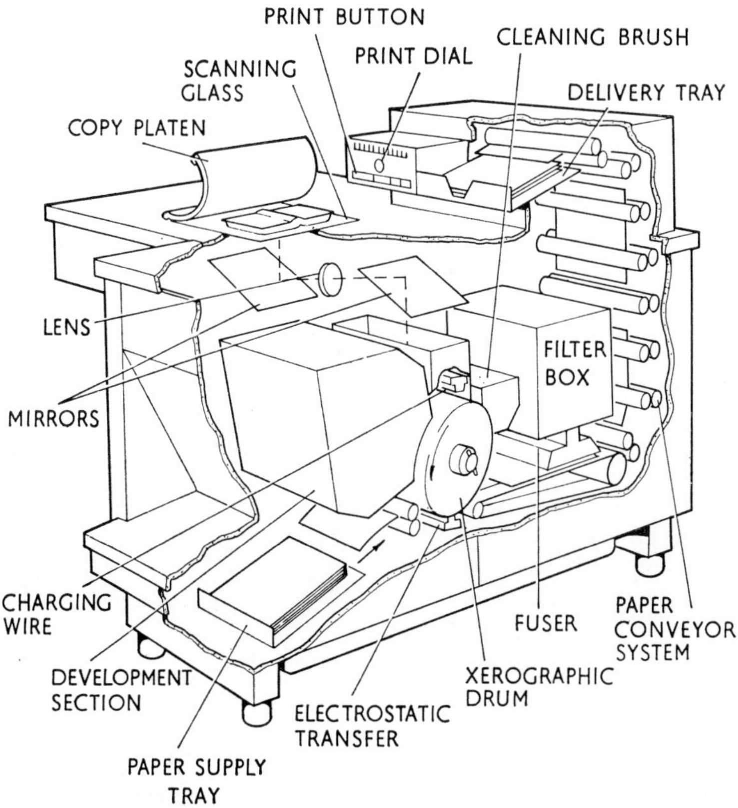

Fig. 21. Schematic diagram typical of a range of free-standing Rank Xerox Office Copier/Duplicators.

duce any distortions, so the mirrors have to be of high quality and precisely located. When they move or oscillate, mechanical tolerances have to be very limited indeed. Sometimes, one well placed mirror with a lens is all that is necessary (see Fig. 19).

There are major differences between the optical systems of machines using sensitized paper and those using normal paper. For

Electrofax paper, the image that emerges from the optical system must be a right-reading positive image on the photo-receptor, because the photo-receptor and the copy are one and the same piece of paper (Fig. 20). In machines using normal paper, the image that emerges from the optical system must appear on the photo-receptor as a wrong-reading image so that when it is transferred by contact to the normal paper, it becomes a right reading copy. (Figs. 21 and 22.)

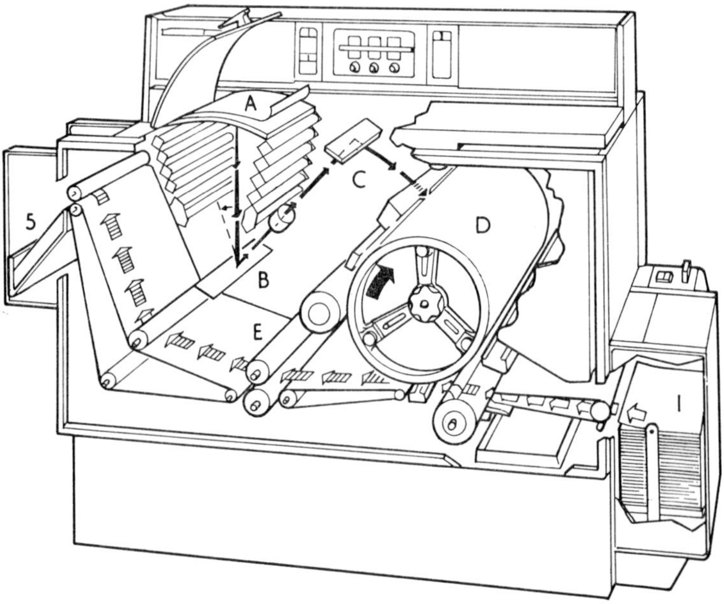

Fig. 22. Schematic diagram of the Rank Xerox 3600 Copier/Duplicator. The original is placed face down on the curved glass platen (A) where two banks of specially developed fluorescent lamps illuminate it from below. The original is scanned by an oscillating mirror (B) which reflects the image and sends it along the optical path through a lens to a fixed mirror (C). From there, it is reflected on to the selenium coated drum (D). The paper on which copies are to be produced is fed in at (1). It is transported along a belt, aligned and brought into contact with the drum (D) where the image is transferred to it. The image is fused by heated roller (E) from where the copy is delivered to the output tray (5).

Path of image to photo-receptor

All rotary electrostatic copiers and/or duplicators currently available use the Xerox or Electrofax method with minor variations in their optical system but arrangements for handling originals and copy paper vary more widely (see page 96).

The following examples from the two major methods illustrate the ways that the image on the original can ultimately reach the photo-receptor.

1. The original can be made to move across a fixed light source. In this case, the lens is fixed and the photo-receptor movement is synchronized with the movement of the original. This is usual with the Electrofax method and some Xerox copiers.
2. The original can remain in one place, while the lights and the lens move and scan the original and the photo-receptor moves in synchronism with the lens and lamps. This method is used in the original copiers sponsored by Xerox Corporation.
3. The original can remain in one place as do the lights and lens while the photo-receptor moves in synchronism with an oscillating mirror which scans the original.

Machines in which the original remains stationery have a wider range of applications than the rotary types. The copying of extracts from books and thick folders, is one such application that comes to mind. This applies to machines using both the Xerox and Electrofax methods where the original is held in this way. Thus we see how the intended applications influence machine design.

Illumination of original

High energy light, that is short wavelength light, is necessary in the right amount to make the photo-receptor switch from its non-conductive electrical state to its conductive state. If this were the only consideration, ultra-violet or white light with a lot of blue in it, would fill the bill. There are, however, some other considerations. There are machine design and commercial considerations, but more important, there are considerations relative to the nature and light response characteristics of the photo-receptor itself.

The selenium photo-receptor needs certain elements to be introduced into it (see page 30) and certain production considerations to be honoured if it is to have even an approximately uniform response to colours in the original. The Electrofax photo-receptor needs certain dyes.

The nature of the light projected is also conditioned by the colours to be expected in the originals to be handled. The Electrofax photo-receptor has peak conductivity with orange light, so projected light should not react with any colours in the original to produce orange light through the lens, which would result in an image voltage so weak that it would hardly attract toner powder. The selenium photo-receptor used in 1961 tended to have a peak response in the blue area of the spectrum so green light was projected.

Photo-response weaknesses in the photo-receptor can be compensated either by changing the nature of the light projected or by incorporating additives to the original photo-receptor. This has been done in the Xerox method and white light, albeit of a special wavelength, is used in the most up-to-date machines. Green light continues to be used in certain other models where a different formulation of the photo-receptor is used. Just any old strip light, or any old household bulb, will not do. There is no point in quoting figures for the ideal light in terms of wavelengths, not only because these figures would be out of date as soon as they were printed, but the nature of the projected light has to be complementary to the nature of the photo-receptors used, and these, in any case, are constantly subject to the results of continuous technical research and development.

Except in the case of the manually operated equipment, where the photo-receptor is imaged through a conventional camera, the lighting arrangements and lens settings cannot be changed to suit the original. It follows therefore that the quality of the final copy is a direct function of the quality of the original. The automated equipment faithfully reproduces the original—including blemishes, paste-up shadows, and fingermarks. What the reprographer used to do to negatives in order to spot out blemishes, pinholes and other defects in the film, is now carried out on the original before processing.

Lens-less equipment

There are a few electrostatic copiers which can image Electrofax papers without a lens system. These machines are restricted in their applications, in so far as they are not able to accept originals which carry images on both sides. The originals, however, do not have to be transparent but they must be translucent. One such lens-less electrostatic copier, which is typical of this group of machines, is the A. B. Dick 610 electrostatic copier shown in Fig. 23.

The 610 operates entirely on a print-through basis, as compared to the normal lens machine which operates on a light projection basis. The original is inserted through a slot in the front of the machine and its lead edge triggers a mechanism which causes the top sheet from the stack of copy paper to be fed into the machine. The sheet of copy paper passes beneath the corona unit and receives a specific electrostatic charge. This operation is accomplished in total darkness in order to prevent accidental removal of the charge from the surface of the copy sheet.

At this stage the charged copy sheet and the original are brought together with the original positioned underneath the copy. The two sheets are held in fixed relationship to each other and are conducted

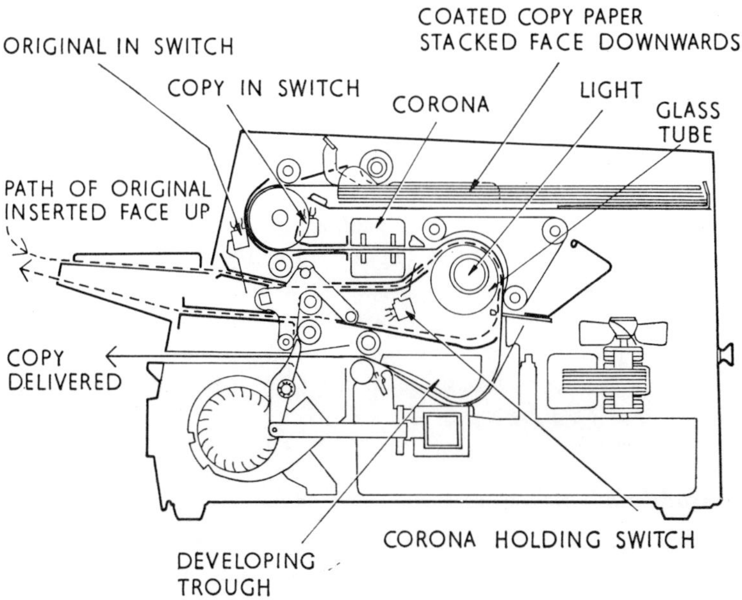

Fig. 23. Schematic diagram of the A. B. Dick 610 lensless Electrostatic Copier.

together part way around the exposure lamp. The light from the lamp passes through the non-image portion of the original and in so doing removes the charge from that portion of the copy sheet. The light which strikes the image is absorbed, thus having little or no affect on the charged portion of the copy sheet which conforms to the area immediately below the image.

During this exposure period the two sheets are moving simultaneously. Immediately following the passage of their leading edges past the exposure lamp they are separated and by means of rollers and other mechanisms the original is returned to the operator. The copy sheet, which carries the electrostatic image of the original, continuous to pass through the machine and into and through the trough which contains the liquid toner. The portion of the copy sheet corresponding to the image on the original absorbs the colouring material in the toner, which results in the electrostatic image becoming a visible image. That portion of the copy sheet which was hit by the light passing through the non-image area of the original has lost its charge and does not absorb colouring material from the toner.

The unit which produces the light needed for exposure is a fluorescent tube of a special design enclosed in a glass cylinder which permits light to pass through but controls the distance from the two sheets of paper to the lamp itself. The fluorescent tube and the glass cylinder enclosing it are fixed in position and do not rotate.

The Electrofax paper is fed into the optical system with its sensitised surface away from the original with which it is going to make contact, so that the image is produced the right way round. A manually operated variable slit aperture allows the operator to control the amount of light necessary to handle variations in the density of the image and the translucence of originals.

The layout of the other types of lens-less machines may vary, but ultimately the relative position of the copy paper and the original has to be as described.

Possible developments

As photo-receptor research results in higher photo-response speeds being achieved shorter exposures will be possible. The incorporation of flash exposure techniques in the design of machines is probable. Both lighting arrangements and machine design will then have to be changed. When this happens very much higher print out speeds will be possible. At present the fastest electrostatic machine is the Rank Xerox 3600 which turns out high quality line copies at the rate of one per second, although much higher speeds are possible with electrographic machines where the input is an electronic alphanumerical signal rather than a hard copy visual communication.

Methods of imaging the photo receptor and the way originals are lit are closely bound up with the nature of the photo receptor itself. In the Electrofax field, chemical research may produce faster photo-response speeds that will warrant the use of flash exposure techniques.

Flexible selenium surfaces could also justify flash exposure methods in machines designed for office use. It might then be necessary to introduce a reusable selenium photo-receptor in the form of a belt rather than a drum to give the surface sufficient time to recover its full photo-response efficiency before being reused.

Undoubtedly, electrostatic print out speeds will increase and inevitably this will cause a redesign of both the machine configuration and the illumination system. If and when bussinesmen opt for the electronic method of transmitting data and information in order to control their affairs, television type screens will provide the electrostatic copier with both the original and the illumination required. In this case, present methods of imaging the photo-receptor and illuminating the original will become obsolete.

CHAPTER VI

DEVELOPING THE IMAGE

Electrostatic and electrographic images are not developed in the photographic sense. The development methods have nothing to do with chemistry but quite a lot to do with physics. To develop an electrostatic or electrographic image, it is necessary to:

1. Choose a pigmented material to be ground to a powder.
2. Arrange for it to be sufficiently fine in texture to provide the resolution quality required.
3. Charge it electrically to the required polarity.
4. Bring it into contact with the latent electrostatic image.
5. Arrange for its constituent particles to coalesce homogenously and to bond into the upper layer of the media which is to carry the image.

These six steps are as simple to understand as they are complex to arrange. They have been arranged in the order dictated by the process sequence. Quite obviously, items 1, 2, and 5 are characteristics of the image-creating material called toner.

Toner characteristics

The grain size of each toner particle used in commercial machines varies from 6–10 microns. For special uses, toners of smaller particle sizes have been produced. A micron is one millionth of a metre and it is difficult to appreciate how small the toner particles really are. It can be said that they are finer than the finest face powder, and leave it at that.

This granule size provides resolution comparable with fine grain photography. In fact, with laboratory techniques, electrostatic prints have been made of live portrait subjects which cannot be distinguished from high grade photoprints of the same subject. The manufacturers have therefore provided reprographers with a resolution quality far in excess of that required for normal commercial reprographic work.

Toner can be made to resist acid for etching purposes and can be made compatible with printing inks. It can be made of a high or low melting point and can be ground to more or less any particle size. Generally, however, users cannot influence the nature of the toner and their safest course is to use the toner recommended for the machine by the suppliers if trouble free operation is to be expected.

Coloured toners. At present the toner is generally available only in black. There are no special difficulties from a technical point of view in producing coloured toners but not all of them can be produced in translucent tints for projection through transparencies. Some are opaque and therefore appear as black when projection illumination systems are used.

The absence of colour toners on the market, arises mainly from commercial considerations. By reason of the size of the plant needed to produce the toner in vast quantities at an economical price, long and continuous runs of one colour are necessary. If the plant had to be used to produce various coloured toners, it would have to be cleaned down of all traces of the previous colour run.

Manufacturers are caught up in a "chicken and the egg" sort of situation. Until the demand for coloured electrostatic prints has crystallized, and until machines are produced in quantity to satisfy this demand so that coloured toners can be used in vast quantities, their introduction is likely to be postponed. On the other hand, until coloured toners are produced economically, the demand for them is not likely to arise. Undoubtedly, coloured toners will eventually appear on the market; the technology required to produce them is known and prototypes of the machines to print in three colours plus black have been produced.

Bonding and fusing ability. So that it shall coalesce and bond into the media, the material chosen is a resin which, when heated, melts momentarily, then becomes solid when plain paper is used. The same effect can be achieved with the concentrated vapour of certain volatile liquids such as trichloroethylene. When this is used, solidification of the image comes about when the vapour evaporates from the toner image it has permeated and made viscous, so that the individual particles, which originally constituted the image, become homogenous. When this method is used, the toner particles tend to swell and very fine line resolution tends to be lost.

When heat is used, the particles will not swell and fine line resolution of the image is better. By reason of the fact that the vapour only attacks the surface of each granule, the resultant image may not be so robust as it is when heat is used to make the powder image permanent. The heat radiations strike to the core of each granule and so the melting process is more complete. For all practical day to day reprographic purposes, however, there is not much difference in image permanency between the two methods.

Vapour fusing is done at room temperature and the method is therefore used when the print-out media—such as acetate—cannot be heated. When, images created by liquid development (see page 74) are fixed, gentle heat is used to accelerate evaporation of the liquid, but this cannot strictly be said to be a vapour fusing method of fixing the image.

Storage and keeping qualities. Except when liquid development is used, the toner must be kept dry; otherwise its very fine granules may matt together with consequent loss of its fine grain resolution characteristics. Properly stored, it will keep for at least a year. It should be kept cool, both in storage and transit; otherwise the granules tend to fuse together; this is sometimes called "cold flow".

Toner that has been subject to warmth before being used, produces a rough feel to the image, especially when solid areas are developed, and a degree of resolution is lost. Under a high power microscope, the electrostatic image surface usually has the smooth appearance of sealing wax. The action that fuses together the individual toner particles is the same action that bonds the image to the media on which it is to be fixed. The minute size of each toner particle makes it easy for the particles on the underside of the image to flow into the surface porosity of the media which is usually sufficiently microscopically pitted to allow the image to be bonded to it.

Charging the toner

The toner has to be charged electrically and has to retain that charge until it has been bonded physically to the copy media. The material chosen is selected so that it can be charged with either a positive or negative charge. The formulation of the toner powder is a commercial secret, its characteristics varying with development methods, applications, type of equipment used and methods of manufacture. All that can be said is that toner consists of carbon black dispersed with a resin.

There are at least five methods of charging the toner, each affecting machine design, the quality of the final print, and to a certain extent, the speed at which the machine can print. The methods are:

Ionic transfer or dry cascade
Electrophoresis or liquid dispersion
Magnetic brush
Fur brush
Aerosol or powder cloud

To understand how the toner gets its negative or positive charge, when used in the cascade method of development, it is necessary to introduce another phenomenon of physics and to explain it in simple terms.

Tribo-electrification methods. The phenomenon is tribo-electrification, which means electrification by friction. When two substances that are non-conductors are rubbed together, they each become electrically charged. The charge each one carries is the reverse of the other. Every day examples of this phenomena are experienced when a nylon shirt is worn under a jacket of similar material. Dry nylon is non-conductive and when the movement of the body causes the two surfaces to rub together, they become oppositely charged and tend to stick together. A woman seldom wears a nylon slip under a dress of similar material, because what happens is that the two materials stick together and wrap themselves round her legs. This was particularly noticeable in the era prior to our female community adopting the miniskirt.

One way of charging the toner is to rub it with a complementary material and, subject to the choice of one of the materials, the toner becomes either negatively or positively charged. What happens to the individual atom of the materials chosen can best be illustrated diagrammatically. Again we have to delve into elementary atomic physics, and consider the electrical characteristics of the atom.

Fig. 11 shows on the left a normally uncharged atom where the electrical charges are equal and therefore balance out and there is no charge. As we have seen (page 48) electrons can be added to, or taken away from the atomic system by the use of a corona discharge arrangement of wires. Electrons can be similarly transferred from one material to another by friction.

Cascade development

When the ionic transfer or dry-cascade method of charging the toner is used, carrier beads are included as a component part of the developer. The carrier bead is about one third of a millimeter in diameter and consists of a glass, steel or Ottawa sand core. This form of core is used so that when coated, it is as near spheroid as possible and will not damage the photo-receptor on rolling over it.

In manufacture, the beads are coated with a special material which is tribo-electrically complementary to the toner material. Thus, when the toner and developer rub together, an exchange of charge takes place; one becoming negatively charged and the other positively charged. It is the choice of coating given to the carrier bead that decides whether the toner shall become negatively or positively charged. The friction between the two is provided by the cascade method itself. Inevitably, the two components are churned up together during the development process, even if they have not been previously churned up in the mixing and packaging or in transit.

The toner is held to the carrier bead by a weak though significant tribo-electric attraction. The charge in the image areas is not only of a reverse polarity, but being stronger, overcomes the toner/carrier bead attraction and so the toner is transferred to the image areas.

The cascade method of development is illustrated in Fig. 24. On the left, just one single carrier bead is considered. Before contacting a charged area of the photo-receptor, it was covered with toner. As it rolls over the imaged area, the image charge detaches the toner by its superior attraction. The developer beads have to be replenished periodically with toner. In the automated machine this is done mechanically by the periodic vibration of a fine screen that separates the toner container from the bucket and chain arrangement that cascades the developer over the surface of the drum as it rotates. The amount of toner dispensed in this way can be pre-set and altered as necessary from time to time, depending upon the density of print areas developed.

Developer replenishment. In manually operated equipment, the developer is replenished with toner by hand. The carrier bead is tinted a light yellow. When toner is added, the developer is darkened through a dark green to almost black. Thus, the operator can adjust the toner concentration by merely observing the overall colour of the developer as it lightens with use. The only way an operator can de-tone developer which has been overtoned, is by making copies.

In practice, it has been found that it is best to ensure that the carrier beads are kept covered with the appropriate amount of toner, i.e. the amount they can take and no more. The toner not only protects the carrier bead coating from dampness when the equipment is not in use, but it seems to preserve its tribo-electric characteristics in much the same way as a keeper to a horseshoe magnet preserves the strength of the magnet.

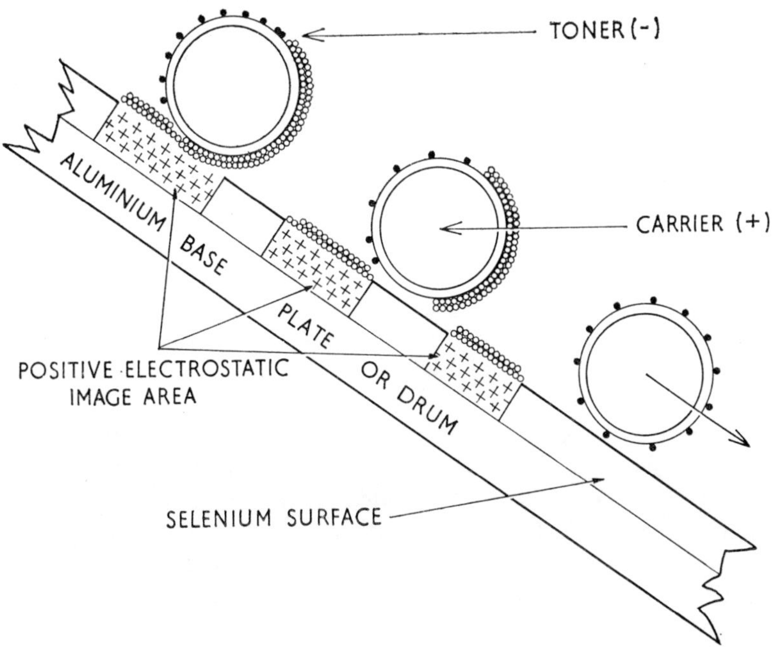

Fig. 24. The cascade method of development. The base which carries the photoreceptor could be a plate or a section of a rotating drum. A succession of toner-coated carrier beads ensures that the charged image areas are coated with toner.

Toning up the developer should be little and often, rather than heavy and occasional. An indication that the developer has reached the end of its useful life is given when the toner, instead of sticking to the beads, deposits itself on the bottom of the developer tray. In this condition, the toner is no longer under the control of the operator and tends, during development, to fly everywhere it is not wanted, including the white areas of the print and so cause background. This applies only to the manually operated equipment. In the automated equipment, the carrier beads are continuously and automatically toned-up.

Curing edge effects. The edge effect noted in connection with the cascade method of development is overcome by arranging for the developer to pass through an electrostatic field generated by an electrode held close to the photo-receptor during development. Whereas this is a standard fitting on manually operated equipment using the cascade method, it has not so far been widely applied to the current automated machines using this method of development. Automated fast duplicators incorporating this modification to the cascade developing method are likely to become commercially available in the early 1970s.

Elfasol method. The Elfasol method of development is basically tribo-electric in that the developer consists of two components each of which becomes charged negative and positive respectively. One of these does not fuse on to the photo-receptor because of its high melting point. The other component has a lower melting point. It therefore, becomes fused on to the image areas or on to the higher melting point component in the non-image areas and is subsequently wiped off at a later stage in the development process, leaving a clean background-free print on a metal offset printing plate. This method of development was originally called the "pull-through double toner" method. The pull-through part refers to the way that the exposed photo-receptor—a metal offset plate—is, in darkness, pulled through a trough of developer. The double toner reference arises because carrier and toner particles are of approximately equal size (five microns).

Electrophoresis or liquid dispersion development

The toner used in the liquid development method, gets it charge during its manufacture. The bulk material is ground down to the requisite particle size in a control agent, such as alkyd resin or linseed oil. Subject to the material selected for the toner, a suitable control agent has to be used at the grinding stage. It is the control agent—either the alkyd resin or linseed oil—that provides the requisite polarity by being absorbed in minute quantities into the toner material during grinding.

The toner material however, has to be chosen so that its specific gravity is as near to the specific gravity of the mixture of insulating liquids which carry it as possible; otherwise, and particularly if it is comparatively coarsely ground, the toner particles will settle and thus the developer will have to be continuously stirred up.

Developer replenishment. Churning of the liquid carrier is also necessary to avoid any part of it being locally depleted of toner when copies are made at the surface. The surface layers of the liquid have to be replenished with toner particles by churning up the whole of the developer. Ultimately, the developer has to be replaced when it becomes depleted of toner. The disposal of used liquid developer could, in certain circumstances, present problems.

In some cases, a fresh charge of toner suspended in its control and fixing agent can be added in order to prolong the useful life of the liquid developer.

Suspending and fixing agents. Sometimes the control agent provides the function of a suspending agent; if not, such an agent has to be added to the liquid. Finally, a fixing agent has also to be added to the developer formulation so that the toner is firmly held to the image carrier or copy medium. The liquid has to be chosen to have a high electrical resistivity greater than 10^{10} ohm-cm so that the rate of destruction of the electrostatic charge is minimized.

It also has to evaporate fairly rapidly if the copy is to emerge reasonably dry. Crude kerosene is often used as the carrier liquid and heaters are often incorporated into the machine to aid evaporation and to dry out the copy. The charged areas of the electrostatic image provide the electrical field force which causes the toner of a reverse polarity contained in suspension to migrate to the image areas.

Curing edge effect. The edge effect noted earlier (see page 54) is also present in this form of development unless an electrode is used to fill up the solid area. Both the churning of the developer and the application of this additional field force can be accomplished by using a roller to apply the toner; the roller itself is the electrode used to counter the hollowing out effect.

Magnetic and fur brush development

The third development method in use commercially is called the magnetic brush method. For convenience, this can be grouped with the fourth—the fur brush method—because the way the toner gets its charge is similar. So far the magnetic brush method has been preferred because it provides better solid area coverage than when fur is used.

The magnetic brush technique is associated with the Bruning Copytron whereas the fur brush method is expected to be used by the Canon NP machine (see page 133). Canon may have second thoughts, because it is known that generally the magnetic brush method renders half-tones of sharper contrast than when the fur brush is used. On the other hand, the research laboratories may have found better fur fibres to overcome this limitation and, who knows, copiers and duplicators may use fur brush development methods in future.

The magnetic brush technique depends upon the fact that iron filings tend to link up into fibres in the presence of a magnetic field and so form the soft bristles of a brush. The magnet can be in the form of a roller with a super soft surface of iron filings covered in toner. The toner is non-metallic so magnetism is not involved in making the toner stick lightly to the "bristles". The toner particles cling at random to the "bristles" and get their charge tribo-electrically as in the other methods previously described.

The brush is made to revolve in a bath of toner which is picked up by the magnetic fibres and carried to the photo-receptor where the electrostatic fields of the image pull the toner particles off the fibres on to the image. In systems using either of these two methods, certain precautions have to be taken. The fibres have to be able to lose their charge to counterbalance the charge carried away by the toner. This can be done by earthing the fibres during the cycle, or combing the bristles or flooding them with corona generated ions.

In the case of the magnetic brush method, the bulk of the developer is conductive. As a result, the image field is modified and solid area coverage is good but, as this system is based on absolute electrical potential rather than adjacent differences in potential, a much more critical control of image exposure is necessary. Line copy quality is high, but continuous tone image quality suffers by reason of the low conductivity of the development system as a whole, does not produce such good solid area coverage. Using conventional toners, the magnetic brush method produces resolution which ranges from ten to forty lines per millimetre.

Aerosol or powder cloud development

The fifth development method is called the aerosol method or powder cloud method. It has a specialized commercial application in connection with xeroradiography. This is not as its name implies, a commercial reprographic process in the strict sense of the term. A xerographic photo-receptor once charged can be selectively discharged by irradiation with X-rays, therefore the process can be used for medical examination, and industrial non-destructive testing purposes by placing the subject to be examined between the X-ray source and the charged photo-receptor. Breaks or dislocations in bone structure or hair-line flaws in castings can be revealed and prints of these can be made on normal unsensitized paper.

Powder cloud methods are used for the development of these exposures and are mentioned, in passing, for the sake of completeness. Most of the work done so far has been with selenium photo-conductors but other photo-conductors can be used. The method has not been used extensively because, in the case of the medical application, X-ray exposures are comparatively long and the doctors are not too keen on subjecting their patients to large doses of X-ray radiation when shorter exposures are possible with photographic film. In the case of the industrial non-destructive testing application, what is required is equipment capable of examining castings delivered on a continuously moving belt and equipment for this purpose has not yet been produced. The industrial application can be used at present only for sample testing purposes, and therefore is not so commercially attractive.

Briefly, this development method uses a gas—normally air—under pressure to convey toner particles through a nozzle in order to create a cloud of toner suspended in air in the vicinity of an electrostatically imaged photo-receptor. By lining the nozzle with the special materials used for coating the carrier bead, the toner is charged to the correct polarity. Other methods of charging the toner that issues from an aerosol nozzle include passing the particles through a corona discharge on their way to the image.

Smoke printing. Smoke printing also comes into this area. It was pioneered by William C. Huebner, but there is no available published technical analysis of the smoke printer later called the Photronic Reproducer by Standard Register Co. The smoke is a mist of ink sprayed through a grid so that it becomes charged. The spray is discharged towards a sheet of paper held against a tin oxide coated glass. This in turn is imaged from a projector lens. The electrostatic charges induce an image charge on the paper which in turn attracts the charged ink mist particles.

Reversal development

All the methods so far described have been concerned with the reproduction of positive originals—black or coloured right-reading images on a plain background which are to be reproduced as black right-reading images on a plain background. The colour of the background depends only on the colour of the media on which the image is created, or to which it is transferred.

Changing the polarity of the toner can cause it to be deposited on the non-image areas of the photo-receptor instead of on the image areas. Thus, a white-on-black photostat or microfilm negative can be reproduced as a black-on-white right-reading image. This is reversal development.

A change in toner polarity can be effected by changing the carrier component with which it is mixed, or, in the case of liquid development methods, to change the conventional liquid for one where the toner has been ground with a different control agent so that the toner particles are reverse charged. The photo-receptor is charged and exposed normally to the original which carries the white image on a black background. When reversal developer is used, the toner is charged to the same polarity as that represented by the black areas of the original. As like charges repel each other, the toner is repelled from the dark areas and settles on the non-charged or image areas shown as white on the original.

Development control techniques

The vast majority of electrostatic copiers and duplicators in use are of the fully automated press-button type. The developer, be it liquid or a dry mixture of powders and carriers, is usually totally enclosed within the machine and, apart from replenishing it from time to time, the user cannot do much with it. He cannot, for example, boost its electrical potential.

Manually operated equipment using the open tray cascade method gives more control over print quality at the development stage because each stage of the image production is separated from all the others and accessible to the operator.

There are two methods of development to choose from. There is the line tray method for line work, where the solid areas are less then $\frac{1}{4} \times \frac{1}{4}$ in. and very fine line resolution is required; and the tone tray method which is used for subjects that include larger solid areas, and continuous tone illustrations. The tone tray method gives a much more intense development of the latent electrostatic image. Faint lines which might not be resolved by the line tray method will thus appear on the final print when the tone tray is used for development purposes. The normal number of cascades is four in the case of the line tray and eight in the case of the tone tray. In certain circumstances the operator is free to increase or decrease the number of cascades in order to improve the quality of the final print.

The strength of the charge on the toner particle is fixed tribo-electrically as a result of the friction between it and the carrier bead, and therefore cannot be increased while in contact with the bead. But once the toner is on the photo-receptor, it can, if necessary, have its negative potential boosted by subjecting the powder image to a negative charge before it is transferred to the media by a positive charge on the media itself. The effect is to increase the difference between the toner potential and the transfer potential and thus more powder is transferred from the photo-receptor to the media. This is a very useful technique to use when faint images on the original result in a low powder density on the photo-receptor which might not otherwise be transferred to the copy.

These modifications to the standard development methods are only possible with the manually operated method of producing electrostatic prints using the dry cascade technique. They cannot be used with either the liquid or magnetic brush development methods.

CHAPTER VII

PREPARING AND HANDLING ORIGINALS

It goes without saying that the reprographic manager who includes electrostatic machines in his print room, or who exercises supervision over electrostatic machines in other parts of the establishment should know the capabilities of his machine. His responsibilities should however go further. Be he called a reprographic manager or preferably a communications manager, he should exercise considerable influence over the choice of the process or machine(s) in the first place. The need for this will become more apparent when considering the check list for examining originals on page 82.

The functional heads of divisions such as the company secretary, chief accountant, director of procurement and so on, should admit that nowadays they are no longer competent to decide which process or machine or combination of these is best suited to the circumstances of their business. Businesses employ transport managers to see to the efficient and economical transport of goods, and packaging managers who see to the packaging of goods; so why should there not be a communications manager who is responsible for the efficient integration of all business communications? Only qualified and experienced reprographers should be considered as candidates for this sort of appointment.

Choice of system and equipment

Visual communications will always be more important than oral communications, if only because the former are permanent and bear witness to the existence of the data and information, whereas the latter are transitory and once forgotten, are lost for ever.

Reprography has now become a very complicated and highly technical business. There are at least twelve different methods of translating or transforming the plans, intentions, directions of management and the language of the computer into meaningful symbols capable of being transmitted to executives, either singly or in multiple copies. Within each of the twelve different methods, there is a myriad of machines, each of which may be highly efficient, economical and convenient for either specialized or general uses. The electronic transmission of hard copy is just around the corner, which will be just another complication and one which will need the reprographer's skills to transform these electronic signals into the language of man. The Videojet and similar systems are examples of this. Having mastered these, the reprographer can take telephonic communications, which are transitory at best, in his stride. The greater always should, and indeed will, absorb the lesser.

The experienced reprographer or communications manager is quite competent to review the present and future visual communications requirements of any business and able to advise the board which is the best combination of processes to be adopted and which are the most useful machines using those processes that should be acquired. The best combination is one which will cater for as near 100 per cent of the requirements at present and in the foreseeable future, as practicable. To approach too near the 100 per cent ideal may, however, entail the serious under-utilization of expensive equipment. There are, after all, plenty of professional reprographic service agencies to cope with the odd 5 or 10 per cent of the company's work which cannot economically be done by the inplant establishment of reprographic machines.

Once the type, nature and size of originals received by or created within the business have been reviewed both as they are now and as they are likely to be in the future, the reprographic requirements can be decided for use on either a centralized or a decentralized basis. The daily loading factor will decide the number of machines to be used. An analysis of the number of originals made by location will help to decide whether centralized, decentralized or both methods should be adopted.

First and second generation copies

The creation of first generation copies on various media presents no problem for the reprographic processes on the market at this present time. Not all these reprographic processes, however, can make second generation copies from the copies made at the first generation stage, unless special materials are involved. The diazo process is an example of this. If a diazo copy is made on translucent material, the copy can also be used to make further copies by the

E.R.—F

same process. If normal diazo paper is used, this is not possible because normal diazo paper is opaque to the ultra violet light that this process needs. Similarly, a copy made by the offset litho or stencil process cannot itself be used to make further offset litho or stencil copies; another master has to be created from the first generation copy by photographic, electronic or manual methods.

With electrostatic copiers and duplicators, the user is not so restricted. Most of these machines make second generation copies from first generation copies because reflex methods of illumination are used. Electrostatic copies made on some zinc oxide coated papers can also be used as a short run master for offset duplication purposes. Electrostatic machines which use unsensitized normal office stationery give this additional facility merely by feeding direct-image paper offset masters into the machine. From an office manager's point of view, the facility for making second generation copies is important because very often the originator does not anticipate that the recipient will require extra departmental copies.

Examination of originals

A "master" is defined on page 10 as that which any particular reprographic machine needs from which to generate further copies. So far as automated electrostatic copiers are concerned, the original is the master. Just as the characteristics of the master govern the quality of the copies made on offset, stencil and spirit machines; so the quality, type and nature of the original govern the copy quality produced by electrostatic copiers and duplicators, all other factors being equal.

When swiftly examining an original presented for processing electrostatically, the user should ask himself or herself the followng questions. These questions are also useful when appraising the nature and type of originals created in any business prior to deciding which type of process and which type of machine is likely to be of the greatest value to the company. Provided that the copies are not expensive, it is obviously better to settle for the most versatile and flexible machine the company can afford.

Here is the check list:

1. Is the original thin, thick, stiff, opaque?
2. Is there good contrast between the image and the background tone of the media which carries it?

3. Is the image black, coloured, or a pale grey, or does it include solid areas larger than $\frac{1}{4} \times \frac{1}{4}$ in.?
4. Does the image include weak or broken characters and/or any broken lines which should be continuous?
5. Is the original a book, or a file from which extracts are to be copied, or is it a solid object—like a slice of bread?
6. Is the original translucent like airmail paper and even if it is not, does the reverse-side image show through?
7. Is it too large for the throat of the rotary machine or is it too large for the copy panel of the flat bed type of machine?
8. Is it a master document from which information or data has to be deleted before the copies are made, or does additional information, which is not on the original, have to be inserted on the copies?
9. Does the original carry dirt, doodles or notes which have to be excluded on the copy?
10. Does the original consist of a set of library or index cards which have to be copied or which have to have their headings listed?

The notes that follow assume that a user will have a significant amount of work to be done under each of these headings. It is only by making this assumption that machine characteristics and operator contribution can be realistically discussed. If the machine is required only to copy good contrast black and white A4 size images which are carried on a good quality paper, practically any of the 40-odd electrostatic machines on the market will do. If, however, the requirement is an integrated visual communication system which is both efficient and economical, the choice is more restricted.

Having acquired an electrostatic copier/duplicator without having checked its capability against the check list, the user is stuck with it and all its limitations. That is why it is so important to define the job to be done within an integrated communications system before any particular machine is acquired.

Unalterable characteristics

Items 1 to 3 are generally outside the control of the reprographer but are within the responsibility of those who select the machine in the first place. For example, very thin originals lacking any substance may, with certain rotary machines, be difficult to feed in. If the original is stiff, like a strip index panel, it cannot be fed into a rotary type machine; a flat bed machine is necessary. If the original is thick, the input throat of a rotary machine designed only to take single sheets of paper, may not be large enough.

The reprographer can do very little about the contrast between the image and the background tone of the media which carries it. If the image is black on a dark red background, then it is doubtful whether any of the electrostatic methods or machines will cope with it. Fortunately, this does not often arise. All the reprographer can do is to warn the originator that the results will be sub-standard if the contrast between the image and the background is poor.

The human eye can read a blue image against a white background, but it is necessary to look at an original through the eyes of the photoreceptor. Such an image may appear to have good contrast but is very difficult to resolve if white light is used for illumination purposes. This is why green light is used to resolve blues electrostatically. The zinc oxide based process handles blue quite well but a yellowish or orange image is difficult to reproduce.

It could be said that these considerations are not very important because both methods can cope with normal colours. The tone and density of the image itself is, however, rather more important. If the image is a pale grey, for example, the toner concentration may have to be increased, so that the copy shows a reasonably black image on a light grey background. A grey image can be blackened up by backing the original with black paper. In this case, the reprographer can only warn the originator of the quality to be expected before the expense of making the copies is entailed. Should the original contain any solid areas larger than $\frac{1}{4} \times \frac{1}{4}$ in. *and* reproduction of these areas as solid black on the copies is vital, then the process and machine capabilities should be checked on this score before the machine is acquired.

Controllable characteristics

The other items in the check list are more or less under the control of the user. Action can be taken to help the machine and sometimes, indeed, to make the copy of better quality than the original.

Weak or broken lines. The weak or broken characters or any lines which should appear on the copy as strong and full characters or continuous lines, should be reinforced with a fine black ball-point pen. If this is not done, the electrostatic process will tend to heighten the contrast between the strong and the weak characters.

Bulky originals. The rotary type machines are not suitable for copying from books, files, folders, etc. Most flat bed types will, however, cope with it. The amount of such work to be done should be carefully noted, because some flat bed machines are produced in two models. One can cope with single sheets while the other considerably more expensive model is required if thick originals are to be processed. The flat bed machines which are designed so that the original does not have to move during processing, are best for this sort of work, if only because they are cheaper to acquire. This is another example of the need to check the requirements before any particular machine is installed.

The example mentioned in the check list, "like a slice of bread", was not given for amusement only. Slices of bread and indeed half loaves, have been copied by flat bed machines which are designed to permit the original to be stationary for the purpose of recording the texture of the bread for sample testing purposes. If solid objects are to be processed the optical characteristics of the machine should have the necessary depth of field incorporated at the design stage.

Image show-through. An original that is so translucent that the reverse-side image shows through should be backed by a sheet of black paper and the two sheets presented to the electrostatic copier. The procedure is repeated with the reverse side so that a single copy is made of both sides of an original which would otherwise be unintelligible if processed normally.

This technique increases the background tone slightly and in this respect gives a somewhat sub-standard copy. This is preferable, however, to the virtually indecipherable copy that would otherwise be produced.

The black paper backing can also be used to blacken up weak typescript, while white paper or card can be placed on the original to mask areas which are to be deleted.

Size of original. Provided that the original can be folded, it is quite possible to produce copies on a flat bed machine section by section. The original should be folded so that on each copy there is a small overlap. One of the copies can be guillotined close to the image area so that the images can be overlapped and the two copies stuck together. With rotary machine, this is only possible where the throat of the machine and the internal mechanism is such that a folded original can be processed in this way.

Some machines on the market use a roll of copy paper which is cut off to the required length of the copy. These machines can process only large sized originals in one dimension, because although the requisite length can be cut off, the roll of paper is of a finite width. There are machines which offer step reduction facilities where large size image areas have to be reduced to the size of the media that the machine is designed to use.

Adding and deleting matter. This is a system documentation application for which both rotary and flat bed machines can be used, but the rotary machines are more restricted. Suitable masks, overlays and templates must be available and the machine must be designed to permit the use of such optional and additional accessories. The need for this kind of application, therefore, has to be anticipated before the copier or duplicator is acquired.

The original has to be cleaned up before processing. Whereas faint pencil marks can sometimes be rubbed out, dirt or ball-point pen doodles cannot be treated in this way. In such cases one or other of the recommended white opaquing compounds should be used to obliterate the defacing images. These compounds should also be used around the edges of any illustrations which have been pasted on to an original.

Copying index cards and listing headings. Where originals have to be copied on to similar materials, machine capability must be checked to see whether it can process unsensitized or plain card stock.

It is possible with some machines to have a template which will permit two or four cards to be processed in one cycle. In this case, A4 size card stock is used on machines designed to accept this sort of copy media and after processing the original in either two up or four up format, the copies are guillotined for separation purposes. Thus the cost of producing the cards is reduced to 20 or 25 per cent of the cost of making one copy.

If the headings of each card, such as the name and address, are required to be listed, a suitable template together with a heading overlay sheet is used so that the cards can be shingled on the object platen of a flat bed machine.

Preparation of originals

Modern copiers do not require originals to be created in any special way: they will faithfully copy what is presented. Nevertheless there are a few points to bear in mind if first class results are to be expected.

Image tone of typewritten matter. It is advisable to use one-time paper carbon ribbons in typewriters and accounting machines to facilitate the production of a truly black copy.

Matching in of addresses and any other additional information can be made undetectable from the information that has been copied. Naturally this assumes that the machines are efficient, properly maintained, serviced and operated. This is the standard that the best machines offer. The advantage of using a one-time paper carbon ribbon is that it cannot be used twice and therefore grey faded images are never created. If these ribbons cannot be afforded, then the originators should not expect first class copy.

The better the contrast between the image and the background which carries it, the better the copies. Good quality white bond paper should be used. Errors can be obliterated with white opaquing fluid and overtyped. Even efficient erasers tend to leave a smudge and bad ones may leave a hole in the paper. An efficient electrostatic copier which is designed to resolve faint images, will naturally copy faint smudges.

Pasting-up. Purified rubber or latex solutions that do not cause the paper to wrinkle are best for pasting in illustrations, texts or diagrammatic insertions. A wrinkled original will include shadow images of the wrinkles.

The edges of pasted-up material tend to cast a shadow and the image of the insert on the copy shows a faint outline. It is best either to make an honest outline of the insert by ruling around the edge with black ball-point pen or to fill up the ridge with white opaquing fluid so that it is a slope and not a ridge.

Tabulated matter. It is best not to use a typewriter to produce ruled lines in tabulated matter. Unless the typewriter is very well adjusted, even horizontal lines look jagged. Vertical lines are a nuisance due to the need to reset the original in the typewriter. A fine black ball-point pen used on a hard surface with a ruler, provides tabulation lines just as quickly and usually produces a better result.

A tabulation is often created on pre-ruled forms but the copy is required without the rules. The figures themselves provide the columnar effect. This looks neater, especially when the columns are close together. The problem is to remove the pre-ruled lines which clutter up the format. All that is necessary is to take a sheet of clear acetate or similar foil, place it over the completed tabulations and run a thin line of white opaquing fluid on the foil over the tabulation lines. The original with the masking overlay in register is presented to the copier and the copies are produced without tabulation lines.

Engineering drawings and microfilm. Large engineering drawings are often reduced to 16 mm or 35 mm microfilm which can be presented to specially designed electrostatic copiers capable of accepting originals of this nature in roll form, or mounted in aperture cards or in microfiche form. These are specialized applications, and in general work such as this, it is sufficient to say that there are British Standards being laid down for both the preparation of the drawings in the first place and for regulating the quality of the microfilm negative or positive from which these industrial electrostatic copiers work. The British Standards so far laid down are BS 4210, BS 308 Appendix 1 and BS 4187.

Manually-operated machines

All that has been said about presenting originals to automated electrostatic copiers and duplicators applies equally well to the manually operated electrostatic offset litho plate makers. The operators of these machines have greater scope for improving quality of sub-standard originals. They can vary the lens aperture, vary the exposure, effect reductions and tidy up the copy by removing blemishes before the powder is finally fused and thus made permanent. Additionally, they can alter the position of the lamps which illuminate the original.

The action of repositioning the lamps can eliminate or create shadows. If the shadows cast, for example, by the overlapping strips of card or paper in a strip index panel are to removed on the copy, an adjustment of the lamps can have quite a startling effect. If the direction of the light can be made as vertical as possible to the ridged surface, the shadows are much less pronounced than if the direction of the light is across the ridges. The operator of manually controlled equipment can also turn the original around in the copy board.

CHAPTER VIII

PRINT-OUT MEDIA

Having understood something of the in-put to electrostatic copiers and duplicators, and having seen how the process works, the next stage to consider is the end product, how important it is and what it is used for in business and commerce. What emerges from these machines is a flexible two-dimensional visual communication which is used to convey information. This description includes ordinary and special papers and offset litho masters.

Modern electrostatic copying and duplicating machines may not be suitable for the production of unsensitized transparent or translucent copies and cannot produce stencil or spirit duplicating masters. In 1957, however, an optional accessory introduced for use with the first manually operated electrostatic offset master maker was capable of producing very good spirit duplicating masters which could print line drawings, typescript and artwork including solid areas and half tones. The quality of the spirit duplicated copies was restricted only by the limitations of the spirit duplicating process itself. The use of this accessory was not encouraged by the suppliers of spirit duplicating machines and its production was abandoned.

Some of the manually operated machines currently on the market can produce printed circuits, rubber stereos for carton printing and cels for film cartoon production and can also be used for printing on cloth and for other specialized applications. The main concern of this chapter is, however, to study the bulk of the usual visual communications found in business and which are produced by automated copiers and duplicators—usually on paper.

The characteristics of the paper that is used have important influences on the creation of originals, the design of machines to copy them and the methods of transferring the images to the other papers which we call copies. Once the paper is made, its characteristics cannot be easily changed, if at all. An opaque sheet of paper can be made translucent to a certain degree, by oiling, preferably with a light volatile oil. This treatment sometimes allows otherwise opaque originals to be made suitable for processing on diazo copiers, but care should be taken to ensure that the oiling of the original does not adversely affect the image. Papers can be stretched and coated but not very much more can be done once the paper has been made. The desired characteristics are built in at the paper making stage.

What paper is made from

All paper is made from cellulose which is a vegetable product. To a varying extent, it occurs in cotton, flax, hemp, jute, ramie, wood, manila, sisal, straw, bamboo, bagasse, maize stalks and esparto grass. The amount of cellulose that is available in these raw materials is high in the case of cotton and reduces as the list lengthens. Those with the most cellulose are more expensive to procure but need less pre-paper making processing, while those with less cellulose may be cheaper to acquire but need more processing. In the final analysis a balance has to be struck between the properties of the cellulose fibres, the nature and quality of the paper to be made and the economic factors associated with production costs, profit margins and market price. The paper maker usually mixes the fibres to bring these factors into balance.

The different raw materials mentioned produce fibres of differing lengths and tubular dimensions. Each of the characteristics are of value according to the kind of paper to be made and the uses to which it is to be put. These range from bank note paper to cigarette paper through envelope and wrapping papers and news print, to high quality writing and blotting papers. The reader is referred to the listed text book for further technical detail regarding fibres.

Since cotton and, to a certain extent, flax is in such high demand for the textile industry, the paper maker has to make do with textile factory rejects in the form of rags and offcuts at one end of the scale, to heavily soiled, worn and coloured rags at the other end of the scale, if he wishes to use cotton fibres in his papermaking process.

When wood is chosen as the raw material, the types of wood used range through spruce and pine to the hard-woods from trees with broad leaves. Man-made fibres are also used in the production of special types of paper such as those used as filter media for both liquids and gases. As 97 per cent of the world's paper is made from wood, this raw material has been chosen as the example to describe in outline how paper is made.

How paper is made

The first step is to make a wood pulp with the object of ending up with a slurry containing cellulose fibres and as little as possible of any other substance. One of these other substances is lignin, a non-cellular substance which is made while the tree grows and which, for want of a better lay description, cements the wood fibres together in the tree. The separation of the fibres and the lignin can be done by mechanical or chemical pulping methods. Mechanical pulping is less complex and therefore less expensive than chemical pulping, but produces a lower quality of paper such as that used as newsprint. The mechanical method employs an abrasive stage where the logs are shredded by friction in order to separate out the fibres and then pressurized steam is used to dissolve the lignin which softens at about 163°C. It is later filtered out of the pulp.

The chemical method requires the logs to be pulverized into chips about two inches long. These chips are then packed into giant-sized pressure cookers with chemicals which digest and dissolve out the lignin. The resulting solution is washed to remove the chemicals, which are subsequently recovered, with the result that the cellulose fibres are left suspended in water. Different chemicals and different cooking temperatures produce pulps of varying types and colour.

Any cleaning and bleaching that may be necessary, takes place at the pulping stage. Once a mixture of cellulose fibres and water of the required purity is achieved, the water is drained off, and the residue compressed, baled and despatched to the paper mills.

At the paper mill, the cellulose fibres have to be reconstituted as a liquid mixture. Originally, the dried pulp was beaten in a stream of water, but nowadays, refiners are used to separate out the individual fibres in a way very similar to the way a housewife uses her liquidizer. The object of this is to prepare the paper stock consisting of very short fibres which range from 0·20 to 0·12 in. long.

The stock is further prepared by adding various substances to adjust the quality of the final paper product. Resins, waxes, mineral fillers, dyes and starches may be added and finally the prepared stock has to be cleaned, strained and have its acidity or alkalinity adjusted, before the paper making process proper can begin.

The stock is fed into a header tank, thence to a breast box and finally under a slice, which is, in effect, a gap or slit. This spreads the slurry on to a wire mesh moving platform, in the form of an endless moving belt. On this wire platform, the water in the slurry drains away and the cellulose fibres begin to settle as it is moved forward.

As more water drains away, a sheet of paper gradually forms over about 50 ft. of travel on the moving wire platform.

Rollers, operating below and just above the wet web of paper, help drain more water away. It is at this stage that watermarks are made, if required. Later in the progress of the wet sheet it is supported by a felt couch which is similar to the wire platform and moves synchronously with it. At this stage, more water is drawn off by suction. When the mixture is from 20 to 30 per cent solid matter and 80 to 70 per cent water, it will support its own weight. Thereafter, more water is pressed out of the mixture and the web of paper can be dried to the correct degree of humidity. It can then be calendered, polished, tub-sized (if this has not already been done at the pulping stage), friction glazed, embossed, corrugated, slit and reeled. Special coatings can be subsequently applied, if this has not been part of the dry end of the process where glazing and calendering was carried out.

What has been described, in a very general way is the method used in the Fourdrinier process of paper making.

Storage and use of paper

Unless paper has a built-in humidity or water content of about 4 per cent or more, it is very stiff and awkward for automated electrostatic machines to handle. Paper left about in high humidity conditions can take up moisture and in reverse conditions can be dried out. Bearing in mind that static electricity can be dispersed by humidity, it is best to use the recommended papers for the particular electrostatic machine. They have been packed in proper moisture proof packets to prevent deterioration in transit or storage and are in optimum condition.

There are other considerations concerning the use of paper in electrostatic copying and duplicating machines. These include the correct guillotining of the cut sheets, the grain and the roughness of the paper used for making copies. The guillotine knife edge must be dust free and not burred. Paper cut by a defective blade causes sheet-to-sheet adhesion in the automatic feed mechanism. Mechanical adhesion between the sheets of a ream of paper can also be caused by the roughness of the paper and may result in two sheets of paper being fed into the machine and causing a paper jam. The most modern copier/duplicators have a double-sheet feed detector which operates a mechanism designed to reject the extra sheet.

Paper grain

The grain of the paper is also important. Paper is stiffer *along* the line of the grain and will bend or curl more easily *across* the line of grain. Paper grain arises by reason of the direction in which the fibres lay on the wire platform and become compacted during the paper making process. Thus a long-grain sheet has fibres which lay in the direction in which the paper was made on the paper machine. This is termed "machine direction" or MD. The term CD describes a sheet of paper which is cut so that the long edge of the paper is cut cross direction or CD which means a short grain paper sheet.

It is easy to discover which way the grain lies in any given sheet by carrying out the following simple test.

1. Cut a strip of paper about an inch wide from the long side of the sheet and mark it for subsequent identification.
2. Cut a similar strip from the short side of the sheet and again mark it. Make both samples the same length.
3. Place the narrow edges together, hold them at one end and suspend them horizontally.
4. Note which of the two samples separates at the free end. If there is no separation, then turn the samples over. The sample that droops less, is the long grain sample.

Where the design of the machine requires the paper to be fed up to a stop for registration purposes, it is obviously better to use a long grain sheet because it is stiffer in this direction and not so likely to curve over the stop. When the paper is required to curve snugly over a drum, then it is obviously better to use a cross grain sheet. It is in these respects that machine designers are in a dilemma. Ideally, they require both characteristics to be present simultaneously, so a compromise has to be made. This is one more reason why only the recommended papers should be used in copying and duplicating machines.

Where the paper feed mechanism uses suction to transport the paper prior to being fed into the duplicating mechanism, or where the design allows the copy paper to be in roll or belt form, paper grain considerations are of lesser importance.

Desirable characteristics for uncoated stationery

For machines which use normal uncoated business stationery, the paper should conform as closely as possible to the following characteristics:

Substance	80 ± 4 gm. per sq. m.
Thickness	0·0038 ± 0·0002 in.
Moisture content	4·7 ± 0·5 per cent moisture by weight
Sizing	25 ± 5 gm. per sq. m.

There are many other characteristics such as stiffness, roughness, brightness, opacity, colour permanence, acidity, porosity, alkalinity, electrical conductivity, tear, burst and curl but they concern only the professional paper maker.

Sensitized papers

The base of sensitized papers has to be electrically conductive so that electrostatic charges can be conducted away from the non-image areas, where they would attract toner and cause background. The base paper has to be made to absorb some of the photo-conductive binder layer for bonding purposes but if it absorbs too much, the electrical characteristics of the whole system are adversely affected. A "solvent holdout" barrier layer is used to prevent excessive penetration of the coating into the base paper.

Zinc oxide crystals are photo-conductive only when they are dry. The zinc oxide therefore has to be contained in a resin binder that prevents ambient humidity from adversely affecting the photo-conductivity of the zinc oxide. The binder layer must be chosen so that it acts as a good insulator in the dark, thus enabling the photo-conductor paper to be charged. It must not interfere with the dye sensitizing of the layer, which would prevent the charges in the white areas leaking away during the exposure stage.

Zinc oxide itself responds conductively only to certain wavelengths of light (see Fig. 7). Dyes have to be added to extend the colour response of the photo-conductor over the spectral range desired. The type of light used to illuminate the original and the colours likely to be encountered in the originals to be processed affect the combination of dyes that have to be incorporated in the binder layer. Finally, the ratio of zinc oxide to binder layer also affects the performance of these types of photo-receptor.

Characteristics of zinc oxide papers

Generally, zinc oxide papers conform to the following outline specification:

Base paper substance	18 – 50 lb per 3,000 sq. ft.
Coating substance	20 – 25 lb per 3,000 sq. ft.
Coating thickness	0·5 – 0·75 mil (thousandths of an inch)
Overall thickness	2·5 – 3·75 mil
Zinc oxide particle size	0·25 – 0·35 micron

The ball milling of the zinc oxide down to the mean particle diameter size shown has to be done carefully. If the crystals are damaged, charge acceptance and the photographic speed of the photo-receptor may be adversely affected.

The material chosen for the binders is critical. Whereas silicon resins permit a higher level of apparent surface voltage and produce somewhat better results over all than the Pliolite 5–7 resin binders, the former has proved to be too expensive. The carboxyl groups tend to increase charge acceptance, whereas the hydroxyl and chloride groups improve photo-sensitivity.

The usual method of applying the layer is by solvent coating but development effort is being directed at the aqueous coating system. Recently it has been noted that a combination of Acryloid B-66 and Amberlac 292X forms a particularly good binder system. The degree of dispersion of the zinc oxide crystals in the binder affects photographic sensitivity and speed. Too sparsely dispersed zinc oxide crystals result in inadequate grain to grain contact nad reduced photo-conductivity.

If these sensitized papers are exposed to light for too long before being used, their electrical characteristics deteriorate with the result that more background than normal appears on the final print. By experience, exposure settings can be selected so that the affect of pre-exposure to light is minimized. As further research and development takes place, it is to be anticipated that coatings will be improved and the hazards of pre-exposure minimized. The electrical conductivity in varying degrees of ambient humidity will be stabilized and papers will become lighter in weight and more translucent.

Limitations of machine design

It is not the print-out media, be it normal or electrostatically sensitized, that restricts the usefulness of the print out in business and commerce. Generally, both normal and zinc oxide coated papers may be used in an integrated communication system. The nature and range of applications is limited by the design of the machines that use the papers. Some of the latest machines can print images on both sides of the paper by running it through twice. Significant savings in the cost of zinc-oxide papers arises, but this is not so important where normal paper is used for copying purposes. The most important benefit of this application, lies not so much in the potentially reduced material cost per copy, but in the reduction of the bulk and weight of paper that needs to be transmitted and stored. By running the paper through twice, the savings in plain paper material costs in Britain is only about 0·3*d.* This is a relatively small part of the total metered cost which, at best, runs out at about 0·9*d.* per copy. The costs of transmission of bulky documents can, however, be significant, especially when it is noted that, generally speaking, zinc-oxide papers weigh more than normal paper.

When processing cheap, direct image offset masters on automated machines which otherwise would use ordinary paper, no special attention is required to the master before being processed. When the longer run paper offset masters are imaged on either the manually operated electrostatic master making equipment, or when imaged by an automated electrostatic copier, it is important to dust off the excess dried china clay powder with which some of these masters are finished. This ensures that the powder image fuses on to a solid surface and not a dusty surface, with the result that the image is more permanent and therefore provides longer print runs. Direct image metal masters can be electrostatically imaged, but only by using the manually operated equipment in conjunction with a special master making accessory. Specially sensitized metal offset masters can be imaged using the Kalle process (see page 36).

Unsensitized print-out media such as transparencies, translucencies, printed circuit blanks and rubber stereos for carton printing, cannot as yet be produced on any of the automated equipment on the market using either of the electrostatic methods. The manually operated electrostatic master making equipment can be used for transferring images on to these types of unsensitized media. Those machines which use liquid development methods are also unable to image the types of print-out media mentioned above. The Bruning Copytron machine which uses magnetic brush development methods can, however, produce translucent sensitized copies suitable for use with the diazo process.

Originally most office copiers using Electrofax papers were cut-sheet fed with sensitized zinc-oxide paper. More recently, web, belt or roll fed methods have become more popular. The advantage of the roll-fed machine lies in the fact that the sensitized paper is automatically cut off to a size which exactly fits the image and wastage is reduced to a minimum. When ordinary paper is used, at about one-sixth the price of sensitized paper, the cut off facility is not so important.

The automatic cut-off facility allows some latitude in the length of originals that these machines can accept. The width, however, is restricted to the width of the roll. Narrow rolls of Electrofax paper are available, so that further material costs can be effected when a large number of small originals have to be processed.

Some reprographic copiers and duplicators can use ordinary paper in roll form. Other specialized machines use a microfilm as an original. They can make copies of any out-of-print books of which a microfilm copy is available or, indeed, of any other microfilmed matter. Some of these machines use individual microfilm frames mounted on aperture cards of a punch card format.

Controlling the output

Electrostatic processes, using either sensitized or unsensitized papers are convenient to use, the copies are easy to make and the machines are versatile in that new documents can be created from existing documents with little, if any, human effort or intervention. The cost of creating these documents, is not prohibitively expensive, ranging in Britain from about 5·5*d.* to 0·9*d.* per A4 size sheet copied or duplicated. It is therefore wise to exercise some control over the amount of paperwork the machines produce.

It is well known that the amount of paper circulating in the average business is reaching unmanageable proportions. Before it becomes completely out of control managements should examine their communication objectives and reprographers should anticipate the possible result of such examination. From a communication point of view, managements' objectives can be expected to include the following:

1. To cease creating hard copy at source and arrange for it to be transmitted electronically to destination where it will be transformed into the language of man.
2. To arrange for all previously created information and data to be made instantaneously available to everyone, everywhere, who needs it for operation purposes.

3. To have the freedom to exchange reprographic equipment without penalty for more or less sophisticated machines as the requirements of an integrated communication system develop or change.
4. To have available processes that offer maximum versatility so that the communication system and its content can change and develop without incurring a prohibitive expense in re-equipping the system with new machines using new processes.
5. Have available reprographic equipment that minimizes the repeated manual transcription of data and information that already exists, in whatever form.

These tentative objectives may represent a counsel of perfection, but that, in itself, is no reason why an integrated communication system should not be designed to approach the agreed objective as closely as possible with the technology available today.

CHAPTER IX

REVIEW OF CURRENT EQUIPMENT

The machines described in this chapter should all be available for some time yet. All the major suppliers have large investments in their production and associated supply business and are not likely to discontinue them before that investment has handsomely paid off. The Verifax machines, for example, appeared about 1952 and are still being supplied. The Rank Xerox 1385 Mastermaker appeared in Britain in 1957 and still has many enthusiastic users in various parts of the world.

Machine modifications, developments and new applications can be gleaned from various trade publications, of which the following are representative in Britain:

- The British Printer
- Business Equipment Digest
- Business Systems and Equipment
- Office Equipment News
- Office Methods and Machines
- Industrial and Scientific Communications
- Graphic Equipment News
- Repro
- Reproduction
- Small Offset Printing
- Print Room

Preliminary survey

The answers to the questions below should facilitate the production of a short list with which to bring under control the problem of choice between, for example, wet and dry machines, desk-top or free standing machines and machines which use roll-fed copy paper or those which make copies on separate sheets of either sensitized or unsensitized paper.

These questions may also be useful in objectively analysing the product features of other types of electrostatic reprographic machines which may appear at a later date.

The check lists of equipment on pages 102 to 117 show how these questions can be applied to various types of machine. With the availability of about forty machines from which to choose, a rational and objective approach to the problem of choice is essential. In the ultimate analysis, a compromise is inevitable because an electrostatic reprographic machine that does every job for everybody in every circumstance with maximum speed, convenience and economy has yet to be produced.

When surveying the field, the intending user should ask himself some questions before venturing out of his office. Apart from those pertinent to his own business, here are some more general ones:

1. Am I proposing to adopt an integrated visual communication system? If so, both the centrally located and the peripherally or remotely sited machines should use the same electrostatic method *and* be capable of handling the systems applications mentioned in Chapter 1.
2. Do I wish to operate a centralized or de-centralized copying and duplicating system, or shall I want to use a combination of both? If so, some, if not all, of the considerations mentioned in (1) are also relevant (see Appendix).
3. What percentage of the whole of my copying and duplicating needs is concerned with the highest quality prestige type print in full colour on expensive paper, and what percentage needs only to be clear, quickly produced, permanent and adequate for decision-taking purposes?
4. Must all machines be desk mounted?
5. Have I the space and conveniently placed power points of the requisite electrical rating for the machines?
6. Should I lock up working capital or should I rent the machines and charge the rental against taxable profits?

The quality of copy produced from any given original can be judged only by a demonstration. Furthermore, the user should have copies made from a specimen set of his own originals. This set should include all various types that he uses or is likely to use and be of good, fair and bad contrast, in all colours he is likely to use. The salesman will offer a suitable original for demonstration purposes, suitable that is for the machine to be demonstrated, but there is nothing quite so reassuring as having your own set of originals reproduced to your own satisfaction.

It is true to say that a low cost per copy plus inconvenience, can equal a higher cost per copy minus inconvenience. The solution to these problems lies in the selection of the reprographic process and the machines associated with it that most clearly attain the user's managerial objectives at the lowest cost, with the shortest contractual commitment for rental, lease and paper supplies which produce an adequate print quality.

Review of equipment

The information given in pages 102–117 has been made as comparative as possible. Where the information is supplied by the manufacturer or distributor, it has not been edited or supplemented. In the other cases, the author has reasonably interpreted the published information made available to him by the suppliers and has had the cost per copy calculated.

The higher cost per copy figure* is calculated on a volume of approximately 7,500 copies per month. The lower figure is based on the most economical method of acquiring the equipment coupled with a high volume usage per month. The user's costs therefore are likely to fall somewhere between the two figures given, depending on local circumstances.

The terms S/S and 1:1 mean that the image on the copy is the same size as the image on the original.

Cost per copy calculations

Of all the information recorded which is supplied by the manufacturers or their agents, the cost per copy claim is the one capable of most variable interpretation. In most cases, the cost per copy varies according to either the usage or the amount of sensitized paper contractually purchased per annum, so the supplier is obliged to make many assumptions in estimating the cost per copy.

The true cost per copy for any given periodic copy volume, using any one of three reprographic processes—electrostatic, offset or stencil—can be calculated from the check list and formulae on page 118. If overheads are to be taken into account, they are apportioned per head of persons employed on an annual basis, because overheads do not normally arise where no one is employed. Once the annual figure has been calculated on a weekly basis, it can be added to each reprographically employed person's wages or salaries. In this case, items P and/or Q would be increased by the overhead

* The costs quoted in this book are in the old currency. A decimal pence conversion table is on page 151.

CHARACTERISTICS OF CURRENTLY AVAILABLE EQUIPMENT

Name	A.M. Copier 2000	A.M. Copier 2100	Apeco Mini-stat	Apeco Super-stat
Rotary or flat-bed type	Rotary	Flat-bed	Rotary	Flat-bed
Development method	Dry magnetic brush	Dry magnetic brush	Toner in liquid dispersant	Toner in liquid dispersant
Optical ratio	Same size	Same size	Same size	Same size
Physical—Length	44 in.	48 in.	$15\frac{1}{2}$ in.	$17\frac{1}{2}$ in.
Height	48 in.	43 in.	$12\frac{1}{4}$ in.	17 in.
Width	49 in.	46 in.	$18\frac{1}{2}$ in.	20 in.
Weight	480 lbs.	730 lbs.	45 lbs.	800 lbs.
Input material (Type) (Size Range) (Bulk)	Thin flexible originals up to 11 in. × 17 in.	3D objects, index panels, ledgers, and single sheets all up to 11·7 in. × 17 in. up to $3\frac{1}{2}$ in. thick	Thin flexible originals up to $8\frac{1}{2}$ in. × 14 in. fed in with copy paper	3D objects, books, index panels and single sheets up to $8\frac{1}{2}$ in. × 14 in.
Max. image area	11 in. × 17 in.	11 in. × 17 in.	$8\frac{1}{2}$ in. × 14 in.	$8\frac{1}{2}$ in. × 14 in.
Copy paper—Type	Zinc oxide in cut sheets	Zinc oxide in cut sheets	Zinc oxide in single sheets	Zinc oxide in single sheets
Weights	Standard	Standard	Standard	Standard
Sizes	Standard	Standard	Standard	Standard
Feed	Manual	Automatic	Manual	Automatic
Capacity	200 sheets	200 sheets	—	250 sheets
Applications and compatibility with other processes:—	Copying Each copy may be used as an offset litho master.	Copying and Duplicating. Each copy may be used as an offset litho master	Copying only	Copying and duplicating

1st Copy delay and Output speed	Not specified 12–14 copies per minute	12 seconds 9 copies per minute	Up to 6 copies per minute	— 6 copies per minute
Electrical Voltages	220–240 volts	220–240 volts	240 volts	240 volts
Frequency	50 cycles	50 cycles	60 cycles	60 cycles
Phase	Single	Single	Single	Single
Current	13 amps.	14·5 amps.	5 amps.	5 amps.
Power	—	—	—	—
Supply terms—Purchase	£1,900	£2,250	£258	£475
Lease	Yes	Yes	—	—
Rent	£7 10s. per month	£12 10s. per month	—	£16 per month
Meter	1·5d. per copy	1·5d. per copy	—	—
Min. contract	£18 15s. per month	£18 15s. per month	—	3 years
Min. charge			—	£576 over 3 years
Cost per copy	3·85d. to 2·60d.	4·01d. to 2·68d.	—	4·10d to 3·73d.
Systems attachments	Not specified	Not specified	Overlays available for adding data to copies	Not specified
Supplier	Addressograph-Multigraph Ltd., Maylands Avenue, Hemel Hempstead, Herts.		Apeco Ltd., Columbus House, North End Rd., Wembley, Middx. Ofrex Ltd., Stephen St., W.1.	
Branches or distributors location	Birmingham, Bristol, Dublin, Leeds, Edinburgh, London, Manchester and other major cities and towns		Not specified	Not specified
After sales service	£108 p.a.	£120 p.a.	Not specified	Not specified
Guarantee Period Contract, etc.	Not specified		Not specified	Not specified
Author's note	Only part of this data was supplied by the Company, the remainder was gathered from leaflets and various sources		Information gathered from various sources	

CHARACTERISTICS OF CURRENTLY AVAILABLE EQUIPMENT

Name	Arcor	Arcor "In Brief"	Agfa-Gevaert Gevafex 50	A-M "500"
Rotary or flat-bed type	Flat-bed	Flat-bed	Flat-bed	Rotary
Development method	Toner in liquid dispersant	Toner in liquid dispersant	Toner in liquid dispersant	Toner in liquid dispersant
Optical ratio	Same size	S/S or 11% reduct.	Same size	Same size
Physical—Length	29 in.	20 in.	31½ in.	26 in.
Height	34 in.	30½ in.	16 in.	16½ in.
Width	38 in.	36 in.	17½ in.	16 in.
Weight	225 lbs.	190 lbs. approx.	112 lbs.	72 lbs.
Input material (Type) (Size Range) (Bulk)	Anything that can be placed face-side against the vertical copy platen up to	Brief size documents	Thin sheets, books and 3D objects	Thin originals
Max. image area	8½ in. × 14 in.	14½ in. × 11¾ in.	8½ in. × 14 in.	11 in. × 14 in.
Copy paper—Type	Zinc oxide in 450 ft. rolls	Zinc oxide in rolls 11¾ in. × 320 ft. or 8½ in. × 450 ft. of standard weight	Zinc oxide in rolls 8½ in. × 492 ft.	Zinc oxide in single sheets
Weights	Standard		Standard	Standard
Sizes	—			
Feed	—			
Capacity	—			100 sheets
Applications and compatibility with other processes:—	Copying and duplicating	Copying and duplicating	Copying and duplicating	Copying

1st Copy delay and output speed	No warm-up 7 copies per minute	No warm-up Up to 10 copies per minute	No warm-up 7 copies per minute	8 seconds 10 copies per minute
Electrical Voltages	220–240 volts	220–240 volts	220 volts	230–240 volts
Frequency	50 cycles	50 cycles	50 cycles	50 cycles
Phase	Single	Single	Single	Single
Current	—	—	5 amps.	12 amps.
Power	2500 watts	2500 watts	—	—
Supply terms—Purchase	£775	£855	£540	£295
Lease	£837*	£1,059 over 3 yrs.	£655 over 3 yrs.	—
Rent	£20 per month†	£64 per month	—	—
Copy rent	—	—	14·3d.–3·5d.	—
Min. contract	† 3 months. * 3 years	3 months	1 year	—
Min. charge		£64	£43	—
Cost per copy	4·0d. to 3·3d.	4·14d. to 4·02d.	6·6d. to 3·5d.	Not quoted
Systems attachments	Not specified	Not specified	Not specified	Not specified
Supplier	Solicitors Law Stationery Soc. Ltd., E.C.4.	Solicitors Law Stationery Soc. Ltd., E.C.4.	Agfa-Gevaert Ltd., Gt. West Road, Brentford, Middx.	Addressograph Multigraph Ltd., Hemel Hempstead
Branches or distributors location	In major cities	In major cities	In major cities	In major cities
After sales service	Free if rented or leased		Provided	Provided
Guarantee period		12 months		Not specified
Contract, etc.		£25 4s. p.a.		
Author's note	Information gathered from leaflet and other sources	Information gathered from various sources	Information gathered from various sources	Information gathered from various sources

CHARACTERISTICS OF CURRENTLY AVAILABLE EQUIPMENT

Name	Copycat 220	Copycat 190	Dennison Standard Model	Dennison High Speed Model
Rotary or flat-bed type	Flat-bed	Flat-bed	Flat-bed	Flat-bed
Development method	Toner and dispersant	Toner and dispersant	Toner in liquid dispersant	Toner in liquid dispersant
Optical ratio	Same size	Same size	Same size	Same size
Physical—Length	$35\frac{3}{4}$ in.	$28\frac{1}{2}$ in.	28 in.	28 in.
Height	11 in.	$10\frac{1}{2}$ in.	45 in.	45 in.
Width	$18\frac{1}{2}$ in.	$17\frac{1}{2}$ in.	21 in.	32 in.
Weight			300 lbs.	470 lbs.
Input material (Type) (Size Range) (Bulk)			Books, index panels, single sheets	Books, index panels, single sheets
Max. image area	No restrictions 10 in. × 15 in.	No restrictions 10 in. × $13\frac{1}{4}$ in.	$8\frac{1}{2}$ in. × 14 in.	$8\frac{1}{2}$ in. × 14 in.
Copy paper—Type	Roll fed	Sheet fed	Zinc oxide in rolls of 688 ft.	Zinc oxide in rolls of 688 ft.
Weights			Standard	Standard
Sizes	8 in., $8\frac{1}{2}$ in. × 10 in. wide	8 in. × 10 in., $8\frac{1}{2}$ in. × 11 in., $8\frac{1}{4}$ in. × $11\frac{3}{4}$ in., 8 in. × 13 in., 10 in. × 14 in.	—	—
Feed			—	—
Capacity			—	—
Applications and compatibility with other processes:—			Copying and duplicating	Copying and duplicating

1st Copy delay and output speed	7 seconds 600 per hour	12 seconds 300 per hour	18 seconds 7 copies per minute	9 seconds 30 copies per minute
Electrical Voltages	220–240 volts	220–240 volts	240 volts	240 volts
Frequency	A/C	A/C	50 cycles	50 cycles
Phase	Single	Single	Single	Single
Current	7 amps.	7 amps.	10 amps.	10 amps.
Power				
Supply terms—Purchase	£579	£409	£990	Rental only
Lease	£13 13s. per month	£9 12s. 6d. per month	—	—
Rent	—	—	£30–£25 per month	£56–£50 per month
Meter			—	—
Min. contract	3 years	3 years	1–24 months	1–24 months
Min. charge				
Cost per copy	2·9d. Quarto	2·9d. Quarto	3·83d. to 3·11d.	4·28d. to 2·72d.
Systems attachments	—	—	Not specified	Not specified
Supplier	Copycat Ltd.	Copycat Ltd.	Sankey Sheldon Ltd., 22 Kingsway, London W.C.2.	
Branches or distributors location	Branches or agents throughout the country	Branches or agents throughout the country	In some large cities	In some large cities
After sales service Guarantee period, Contract, etc.	1 year—spare parts 6 months—labour	1 year—spare parts 6 months—labour	Maint. contract £45 p.a.	All calls and spares free
Author's note	Information and data provided in full by the supplier		Information gathered from various sources	

CHARACTERISTICS OF CURRENTLY AVAILABLE EQUIPMENT

Name	The Dalco 350 Electrostatic Copier	The Dalcostatic	The A.B. Dick 675	The A.B. Dick Model 650
Rotary or flat-bed type	Rotary	Rotary	Flat-bed	Rotary
Development method	Toner in dispersant liquid	Toner in dispersant liquid	Toner in dispersant liquid	Toner in dispersant liquid
Optical ratio	Same size	Same size	Same size	Same size
Physical—Length	27 in.	21 in.	18½ in.	26¾ in.
Height	15 in.	14 in.	11 in.	19 in.
Width	18½ in.	19 in.	35¾ in.	18 in.
Weight		64 lbs.	190 lbs.	123 lbs.
Input material (Type) (Size Range)	Thin flexible originals up to 11¾ in. wide	Thin flexible originals up to 16 in. × 13 in.	Books, index panels, single sheets up to 10 in. × 15 in.	Flexible originals of any length up to 11 in. wide
(Bulk)		Cut sheet feed		
Max. image area	11 in. by any length	Not specified	10 in. by any length	11 in. by any length
Copy paper—Type	Zinc oxide in rolls 8 in., 8½ in., 11 in. Min. supply 2 rolls each 460 ft. costing per 11 in. roll—£16 13s. 4d.	Zinc oxide Details not specified	Zincoxide in 460 ft. rolls 8 in., 8½ in. and 10 in. £8 6s. 8d. per 10 in. roll.	Zinc oxide in 400 ft. rolls 8 in., 8½ in. and 11 in. wide.
Applications and compatibility with other processes:—	Copying only	Copying only	Copying, duplicating & offset master-making	Copying and duplicating

1st Copy delay and output speed	Not specified	6 per minute	1st copy in 10 seconds then 10 per minute	1st copy in 20 seconds then 7 per minute
Electrical Voltages	Not specified	Not specified	220–240 AC	220–240 AC
Frequency			50 cycles	—
Phase			—	—
Current			10 amps.	7 amps.
Power			—	—
Supply terms—Purchase	£395	£245	£579	£555
Lease	—	—	£729 for 3 years	Can be leased
Rent	£434 10s. 0d. p.a.	—	13 gns. per month	—
Meter	—	—	—	—
Min. contract	£181 0s. 10d.	—	Rental—3 months	—
Min. charge	—	—	—	—
Cost per copy	4·03d. to 3·72d.	Less than 3d.	4·34d. to 3·83d.	Not specified
Systems attachments	Not specified	Not specified	Not specided	Special carrier for lightweight and small originals
Supplier	Dalcopy (U.K.) Ltd.	Dalcopy (U.K.) Ltd.	A. B. Dick Co. of G. B. Ltd.	A. B. Dick Co. of G. B. Ltd.
Branches or distributors location	Not specified	Not specified	Branches in major cities	Branches in major cities
After sales service	Not specified	Not specified		
Guarantee period			12 months	12 months
Contract, etc.			£25 p.a.	By quotation
Author's note	Data gathered from leaflet	Data gathered from leaflet	Data gathered from various sources	Data gathered from various sources

CHARACTERISTICS OF CURRENTLY AVAILABLE EQUIPMENT

Name	Minolta Fax Mark I	Minolta Fax Mark II	Minolta Rotary	Pitney-Bowes 250
Rotary or flat-bed type	Flat-bed	Flat-bed	Rotary	Rotary
Development method	Fluid	Fluid	Fluid	Toner in liquid dispersant
Optical ratio	1 : 0·82	1 : 0·73	Same Size	Same size
Physical—Length	$27\frac{1}{2}$ in.	$27\frac{1}{2}$ in.	18 in.	$26\frac{1}{4}$ in.
Height	44 in.	44 in.	14 in.	15 in.
Width	35 in.	33 in.	$18\frac{1}{2}$ in.	$15\frac{3}{8}$ in.
Weight	275 lbs.	290 lbs.	64 lbs.	85 lbs.
Input material (Type) (Size Range) (Bulk)	Any original up to 10 in. × 14 in. Books, etc.	Any original up to $11\frac{1}{4}$ in. × 16 in. Books, etc.	Any original $11\frac{3}{4}$ in. by any length Up to 12 sheet	Thin originals
Max. image area	10 in. × 14 in.	$11\frac{1}{4}$ in. × 16 in.	$11\frac{3}{4}$ in. by any length	$8\frac{1}{2}$ in. by any length
Copy paper—Type	Zinc oxide	Zinc oxide	Zinc oxide	Zinc oxide in rolls $8\frac{1}{4}$ in. × 285 ft. Standard
Weights	90–95 gms.	90–95 gms.	90–95 gms.	
Sizes	A.4	A.4	Cut sheets	
Feed	Auto	Auto	Manual	
Capacity	250	250	Single feed	
Applications and compatibility with other processes:—	Facility to produce masters for dyeline copying and offset duplicating			Copying

1st Copy delay and output speed	27 seconds Every 10 seconds	27 seconds Every 10 seconds	10 seconds 6 per minute	15 seconds 6–7 per minute
Electrical Voltages	220–240 volts	220–240 volts	220–240 volts	230–250 volts
Frequency	50 cycles	50 cycles	50 cycles	50 cycles
Phase	Single	Single	Single	Single
Current	AC	AC	AC	13–15 amps.
Power			220–240 volts	1300 watts
Supply terms—Purchase	£685	£785	£239	£375–£550
Lease	Yes	Yes	Yes	£462 over 3 years
Rent	Yes	Yes	No	£10 per month
Meter	Yes	Yes	No	—
Min. contract	1 month	1 month	1 year	1 year
Min. charge	£16 13s. 4d.	£16 13s. 4d.	—	—
Cost per copy	3·2d.	3·2d.	2·7d.	3·36d. to 3·08d.
Systems attachments				Not specified
Supplier	Anson-Skycopy Ltd., 75–79 Southwark Street, London, S.E.1.			Pitney-Bowes Ltd., The Pinnacles, Harlow, Essex.
Branches or Distributors location	Service depots and stockists in all major towns			In all major cities
After sales service Guarantee period, Contract, etc.	12 months Maintenance agreement			Free if Pitney-Bowes paper is used
Author's note	Information and data provided in full by the supplier			Information gathered from various sources.

CHARACTERISTICS OF CURRENTLY AVAILABLE EQUIPMENT

Name	Remington R.2	Rank Xerox 813	Rank Xerox 660	Rank Xerox 420
Rotary or flat-bed type	Flat-bed	Rotary	Rotary	Flat-bed
Development method	Toner in liquid dispersant	Dry electrostatic cascade	Dry electrostatic cascade	Dry electrostatic cascade
Optical ratio	Same size	6% reduction	Same size	Same xize
Physical—Length	29 in.	28 in. 71·12 cm.	28 in. 71·12 cm.	$45\frac{1}{2}$ in. 115·6 cm.
Height	16 in.	18 in. 45·70 cm.	18 in. 45·70 cm.	41 in. 104·1 cm.
Width	19 in.	20 in. 50·8 cm.	20 in. 50·8 cm.	$45\frac{1}{2}$ in. 115·6 cm.
Weight	135 lbs.	230 lbs.	235 lbs.	685 lbs.
Input material (Type) (Size Range) (Bulk)	Anything up to $\frac{1}{4}$ in. thick Measuring $11\frac{3}{4}$ in. × 21 in.	Thin flexible documents from 9 in. × 14 in. to $5\frac{1}{2}$ in. × $5\frac{1}{2}$ in.	Thin flexible documents from $8\frac{1}{2}$ in. × 13 in. to $5\frac{1}{2}$ in. × $5\frac{1}{2}$ in.	From thin flexible originals to bound books, index panels
Max. image area	11 in. by any length	$8\frac{1}{2}$ in. × 13 in.	8 in. × 13 in.	9 in. × 14 in.
Copy paper—Type	Zinc oxide in rolls of 11·69 in. to $5\frac{1}{2}$ in. × 550 ft.	Normal unsensitized	Normal unsensitized	Normal unsensitized
Weights		21 lbs.	80 gsm.	76 to 175 gsm.
Sizes		8 in. × 10 in.— $8\frac{1}{2}$ in. × 13 in.	8 in. × 10 in.— $8\frac{1}{2}$ in. × 13 in.	8 in. × $5\frac{1}{2}$ in.— 10 in. × 15 in.
Feed		Friction	Friction	Friction
Capacity		200 sheets	200 sheets	200 sheets
Applications and compatibility with other processes:—	Copying and duplicating	Copying and systems uses with masks and overlays	Copying, duplicating and systems uses with masks and overlays	Copying, duplicating and systems uses with masks and overlays

1st Copy delay and output speed	7 seconds 18–30 copies per minute	25 seconds 5 copies per minute	15 seconds 7 copies per minute	90–30 seconds 7 copies per minute
Electrical Voltages	240 volts	90–260 volts	90–260 volts	90–260 volts
Frequency	50 cycles	50 cycles	50 cycles	50 cycles
Phase	Single	Single	Single	Single
Current	8 amps	12 amps	6·9 amps. at 260 volts	11 amps. at 230 volts
Power		1400 watts	1800 watts	2·5 kvA.
Supply terms—Purchase	£545	—	—	—
Lease	£675 over 3 years	—	—	—
Rent	Not available	—	—	—
Meter	—	4·7d. to 3·7d.	4·7d. to 1·5d.	4·1d. to 1·5d.
Min. contract	3 years	6 months	30 days	30 days
Min. charge	—	£12 15s. 10d. per month	£25 per month	£30 per month
Cost per copy	4·24d. to 3·66d.	5·0d. to 4·04d.	4·7d. to 4·0d.	3·97d. to 3·92d.
Systems attachments	Not specified	Flexible overlays and masks available	Flexible overlays and masks available	Overlays, masks and templates available
Supplier	Remington Rand Ltd., Holborn Viaduct, E.C.1.	Rank Xerox Ltd., 338 Euston Road, London N.W.1.		
Branches or distributors location	In all major cities	In all major cities and towns at home and abroad	In all major cities and towns at home and abroad	In all major cities and towns at home and abroad
After sales service Guarantee period, Contract, etc.	Free for 12 months then £25 p.a.	Included in the meter rental charge		
Author's note	Information gathered from leaflets and other sources	Data and information provided by the supplier		

CHARACTERISTICS OF CURRENTLY AVAILABLE EQUIPMENT

Name	Rank Xerox 3600 and Sorter	Rank Xerox Computer Forms Printer	Rank Xerox System 600	Rank Xerox 1824U
Rotary or flat-bed type	Flat-bed	Flat-bed	Flat-bed	Flat-bed
Development method	Dry electrostatic cascade	Dry electrostatic cascade	Dry electrostatic cascade	Dry electrostatic cascade
Optical Ratio	Same size	Same size or 23% reduction	12×, 14·5×, 15× or 16× magnification	14·5× magnification
Physical—Length	98 in.	117 in.	90 in.	38 in.
Height	$45\frac{1}{2}$ in.	$36\frac{1}{2}$ in.	$41\frac{1}{4}$ in.	65 in.
Width	31 in.	31 in.	$52\frac{1}{4}$ in.	34 in.
Weight	600 lbs.	1970 lbs.	1150 lbs.	660 lbs.
Input material (Type) (Size Range) (Bulk)	Flexible originals that conform to the curved platen up to $8\frac{3}{4}$ in. × 14 in.	Punched and perforated continuous business forms $4\frac{1}{2}$ in. × $5\frac{1}{2}$ in.— $15\frac{5}{16}$ in. 15 lb. to 36 lb.	Microfilm aperture cards	Aperture cards negative or positive 35 mm. or 16 mm. roll microfilm
Max. image area	$8\frac{1}{2}$ in. × 13 in.	$8\frac{1}{2}$ in. × 13 in.	On A2 paper	On 18 in. × 24 in.
Copy paper—Type	Normal unsensitized	Pre-punched or plain cut sheets	13 to 20 lb. bond	Normal bond paper, translucent and offset master stock
Weights	20 lbs.	20 lb. bond		
Sizes	8 in. × 10 in.— $8\frac{1}{2}$ in. × 13 in.	10 in. × 8 in.— 13 in. × $8\frac{1}{2}$ in.	A2, A3 and A4	
Feed	Suction	Suction		
Capacity	2,000 sheets	2,000 sheets	200 aperture cards	
Applications and compatibility with other processes:—	Copying, duplicating and systems uses with masks and overlays	Copying, duplicating and sorting. Using masks and overlays	Printing enlarged images of data on microfilm aperture cards	Printing enlargements from aperture cards

1st Copy delay and output speed	Nil on standby 60 copies per minute	Nil on standby 40 copies per minute	2½ minutes 600 per hour	70 seconds 2–5 copes per minute
Electrical Voltages	200–440 volts		280 or 240 volts	220–250 volts
Frequency	50 cycles	50 cycles	50 cycles	50 cycles
Phase	Single	Single	Single	Single or 3
Current	22 amps. at 240 volts		40 amps.	22 amps.
Power		6 kvA	8 kvA	
Supply terms—Purchase	—	—	—	—
Lease	—	—	—	—
Rent	—	—	—	—
Meter	3·7d. to 0·75d.	By quotation depend-ing on volume	By quotation depend-ing on volume	By quotation depend-ing on volume
Min. contract	90 days			
Min. charge	£155 per month			
Cost per copy	4·0d. to 0·90d.			
Systems attachments	Flexible overlays and masks available	Flexible overlays and masks available	Line screens for full and half-tone printing available	
Supplier	Rank Xerox Limited, 338 Euston Road, London, N.W.1			
Branches or distributors location	In all major cities and towns at home and abroad			
After sales service Guarantee period, Contract, etc.	Included in the meter rental charge			
Author's note	Data and information provided by the supplier			

CHARACTERISTICS OF CURRENTLY AVAILABLE EQUIPMENT

Name	SCM Model 55	SCM Model 44	SCM Model 88	Statikon
Rotary or flat-bed type	Flat-bed	Rotary	Rotary	Rotary
Development method	Electrostatic	Electrostatic	Electrostatic	Dry Cascade
Optical ratio	1 to 1	1 to 1	1 to 1	1 to 1
Physical—Length	36 in.	27 in.	27 in.	$26\frac{3}{4}$ in.
Height	35 in.	16 in.	16 in.	$14\frac{1}{2}$ in.
Width	20 in.	18 in.	18 in.	$13\frac{1}{2}$ in.
Weight	279 lbs.	110 lbs.	110 lbs.	97 lbs.
Input material (Type)	Anything which can be placed on the platen	Paper	Paper	Single sheets up to 9 in. × $13\frac{1}{2}$ in.
(Size Range)	up to 10 in. × $15\frac{1}{2}$ in.	Up to 11 in. × 17 in.	Up to 11 in. × 17 in.	
(Bulk)	Any thickness	Single sheet	Single sheet	
Max. image area	9 in. × 14 in.	11 in. × 17 in.	11 in. × 17 in.	8 in. × $12\frac{1}{2}$ in.
Copy paper—Type	Zinc oxide	Zinc oxide	Zinc oxide	Uncoated all purpose paper
Weights	Standard	Standard	Standard	83–100 gsm.
Sizes	Standard sizes	Standard sizes	Standard sizes	$8\frac{1}{2}$ in. × 13 in.
Feed	Up to maximum as above	Up to maximum as above	Up to maximum as above	—
Capacity	500 sheets	250 sheets	250 sheets	100 sheets
Applications and compatibility with other processes:—	Offset paper plates can be made	Offset paper plates can be made	Offset paper plates can be made	Copying and making offset plates, dyeline masters double-sided copies.
1st Copy delay and output speed	8–10 seconds Up to 10 copies per minute	8–10 seconds Up to 8 copies per minute	7 seconds Up to 14 copies per minute	Nil 4 copies per minute

Electrical Voltages	220–250 AC	220–250 AC	220–250 AC	240 volts
Frequency	50 cycles	50 cycles	50 cycles	50 cycles
Phase	Single	Single	Single	Single
Current	10 amps.	8 amps.	8 amps.	13 amps.
Power	2500 watts	2000 watts	2000 watts	—
Supply terms—Purchase	£1750	£475	£575	£540
Lease	Yes	Yes	Yes	£670 over 3 years
Rent	Yes	Yes	Yes	Not available
Meter	Yes	Yes	Yes	—
Min. contract	30 days	30 days	30 days	3 years
Min. charge	Varies according to quantity	Varies according to quantity	Varies according to quantity	
Cost per copy				Down to 1·25d.
Systems attachments	Templates and overlays available for all machines			Not specified
Supplier	S.C.M. (United Kingdom Ltd. 248–250 Tottenham Court Road, London, W.1			Office and Electronic Machines Ltd., London, S.E.1.
Branches or distributors location	London, Birmingham, Manchester and Leeds Direct representation throughout England and Wales			In major cities
After sales service Guarantee period, Contract, etc.	Throughout England and Wales, and by dealers in Scotland 12 months including parts Service contract £28, including parts			1 year Voucher system
Author's note	Information and data provided in full by the supplier			Most of the information provided by supplier

factor. This will provide only an approximate figure for overheads, but it is better than nothing and it allows each user to define an overhead for himself.

Before the check list and formulae can be used, the volume and the nature of the mixture of reprographic work done over (say) a month has to be decided. For example, the daily average volume may be 900 copies which arise from 12 originals from each of which 75 copies may have to be made. This would represent the loading of a low volume user. Alternatively, the average daily loading might consist of 50 originals from each of which either 20 or 150 copies are to be made. Where the daily reprographic requirements have no particular pattern and consist of a mixture of many requirements, the requirements have to be grouped over a month and the costings for each group worked out.

The check list and formulae could be used to construct a table showing the cost per copy at various loadings. Such a table is useful where a variety of work needs to be costed so that the most economical method can be chosen. It could also be used for any one of the six reprographic methods mentioned. Following the check list an example has been fully worked out in order to show how the formulae are applied. The variable items which need to be inserted or assumed before the formulae can be applied are at Serials A, B, C, E, F, G, H, I, J, K, P, Q, S, X, Y and Z. Having substituted the actual figures, applied the formulae and inserted the results in the column, the items printed in italics have only to be added together where applicable to arrive at the cost per copy for the particular example chosen.

COST PER COPY CHECK LIST AND FORMULAE

Example—A low, medium or high volume user of reprographic equipment which produces *X* copies of A4 size per working day of a 5-day week by the various methods listed below. The number of masters made and the number of copies duplicated to be averaged over a month.

Methods reviewed—Offset-litho duplicating from a typed offset paper master or from an electrostatically made master. Stencil duplicating from a typed stencil master or from an electronically cut stencil master. Electrostatic duplicating direct on Electrofax paper or on normal paper.

The figures given in the worked out example represent those of a low volume user. In this case, the machine and labour elements are

Serial	Item	Formulae and Detail
A	Price of machine(s)	Selling price
B	Life of machine(s)	for depreciation
C	Service contract	per annum
D	Annual cost of machine(s)	$(A \div B) + C$
E	Annual rent or lease *or*	
F	All-in meter-rental terms	per 1–3 copies* per 4–10 copies* per 11 plus copies*
G	Minimum contract per annum	for paper or machine
H	No. of working days per annum	52×5
I	No. of masters made/originals used per day	
J	No. of masters made per hour	
K	No. of copies made per master or original	
L	No. of copies made per day	
M	Machine fixed costs per day	$D(\text{or } E) \div H$
N	*Machine fixed costs per copy*	$M \div L$
P	Labour—male per week	Incl. NHI. SET. Pension, etc.
Q	Labour—female per week	Incl. NHI. SET. Pension, etc.
R	*Duplicating labour cost per copy*	$P(\text{or } Q) \div (5 \times L)$
S	No. of hours worked per day	
T	*Master making labour cost per copy*	$(P(\text{or } Q) \div (5 \times I \times K)) \times (I \div (S \times J))$
U	*Master making machinist or key operator*	$Q \div (5 \times L)$ per copy
V	*Master making—materials, cost per copy*	Cost of Master $\times I \div L$
W	*Master making—machine cost per copy*	$M + (X \times I) \div L$
X	*Plain or sensitized copy paper*	per sheet
Y	*Inks and Fluids*	per copy
Z	*Other Consumables*	per copy
	Cost per copy	$(N + R + T + U + V + W + X + Y + Z)$

Notes

* Rental included if the total meter charge is above G monthly figure.

If the minimum monthly contract applies then $N = G \div (H \times L)$ and this becomes a cost per copy element and F does not apply.

considerably under utilized and consequently the cost per copy in some cases is high. The formulae suggested and the methods of tabulating the cost factors are specifically designed to reveal these circumstances. Obviously certain items must be the same for each of the processes if realistic comparisons are to be made for the same job. By varying the other figures in the tabulation and the example, the reader can get some idea of an economic break-even point

PRACTICAL COST PER COPY EXAMPLE

Example—A low-volume user of reprographic equipment which produces 900 copies of A4 size per working day of a 5-day week by the various methods reviewed below. The number of masters made and the number of copies duplicated, have been averaged over a month.

Methods reviewed—

1. Offset-litho duplicating from a typed offset paper master
2. Offset-litho duplicating from an electrostatically made master
3. Stencil duplicating from a typed stencil master
4. Stencil duplicating from an electronically cut stencil master
5. All copies electrostatically produced on Electrofax paper
6. All copies electrostatically produced on normal paper

Serial	Item	Formulae and Detail	1.	2.	3.	4.	5.	6.
A	Price of machine(s)	Selling price	£500	£500	£80	£690	—	—
B	Life of machine(s)	for depreciation	5 yrs	5 yrs	5 yrs	5 yrs	—	—
C	Service contract	per annum	£26	£26	£4	£10	—	—
D	Annual cost of machine(s)	(A ÷ B) + C	£126	£126	£20	£148	—	—
E	Annual rent or lease *or*		—	see W	—	—	£315	£360
F	All-in meter-rental terms	per 1–3 copies } Rental included if the total meter charge is above G monthly figure	—	—	—	—	—	4·1d.
		per 4–10 copies	—	—	—	—	—	2·3d.
		per 11 plus copies	—	—	—	—	—	1·5d.
G	Minimum contract per annum	for paper or machine	—	—	—	—	£315	£360
H	No. of working days per annum	52 × 5	260	260	260	260	260	260
I	No. of masters made/originals used per day		12	12	12	12	12	12
J	No. of masters made per hour		3	see W	3	—	—	—

K	No. of copies made per master or original		75	75	75	75	75	75
L	No. of copies made per day		900	900	900	900	900	900
M	Machine fixed costs per day	D(or E) ÷ H	116d.	116d.	18d.	137d.	291d.	332d.
N	Machine fixed costs per copy	M ÷ L	*0·13d.*	*0·13d.*	*0·02d.*	*0·15d.*	*0·32d.*	see F
P	Labour—male per week	Incl. NHI. SET. Pension, etc.	£20	£20	—	—	—	—
Q	Labour—female per week	Incl. NHI. SET. Pension, etc.	£15	£15	£15	£15	£15	£15
R	Duplicating labour cost per copy	P(or Q) ÷ (5 × L)	*1·06d.*	*1·06d.*	*0·8d.*	*0·8d.*	see U	see U
S	No. of hours worked per day		6	6	6	6	6	6
T	Master making labour cost per copy	(P(or Q) ÷ (5 × I × K)) × (I ÷ (S × J))	*0·53d.*	—	*0·53d.*	*0·03d.**	—	—
U	Master making machinist or key operator	Q ÷ (5 × L) per copy	—	*0·8d.*	—	—	*0·80d.*	*0·80d.*
V	Master making—materials, cost per copy	Cost of Master × I ÷ L	*0·05d.*	*0·02d.*	*0·16d.*	*0·66d.*	—	—
W	Master making machine cost per copy	(M(Col. 5) + (X × I)) ÷ L for X see X–Col. 5	—	*0·35d.*	—	—	—	—
X	Copy paper	per sheet	*0·3d.*	*0·3d.*	*0·3d.*	*0·3d.*	*1·74d.*	*0·3d.*
Y	Inks and fluids	per copy	*0·02d.*	*0·02d.*	*0·02d.*	*0·02d.*	—	—
Z	Other consumables		—	—	—	—	*0·38d.*	see F
	Cost per copy—add italic figures and calculate F (1·66d.) for Col. 6 figure		2·09d.	2·68d.	1·83d.	1·96d.	3·24d.	2·76d.
			or 1·83d.	2·41d.	if females operate the offset machines.			

Note—In Column 6 if the minimum monthly contract applies, N = G ÷ (H × L) = 0·37d. and item F col. 6 does not apply.

* One minute at £15 per week assuming a 6-hour day per copy from one master.

between the processes and discover which is the most economic process for any given set of circumstances.

The figures are intended only to show how the formulae are applied. It is not intended to imply that any one process is necessarily and generally more or less economic than any other. The real value of the check list lies in the columns headed "Item" and "Formulae and Detail" for use with an example that is most closely associated with the actual circumstances with which the reader has to deal.

At item V in column 4 of the worked out example, the master material cost is taken as 4*s*. Cheaper masters could be used if the original consisted of line work only. The last point to note in checking the figures is that the copy paper (Serial X) varies between suppliers and differs according to the method of electrostatic development used. Item C should be increased by the annual cost of spares if these are not included in the service contract, the leasing arrangement, or the annual rental. This additional expense would most likely arise in the case of machines over twelve months old which are heavily loaded.

"Best method" considerations

In the specific example worked the cost per copy can vary from 3·24*d.* to 1·83*d.* At first sight, this variation could give some indication as to which is the best method. At 1·83*d.* per copy (col. 3) the example happens to show that the col. 3 method is the most economical. If, however, the copy volume was less, the col. 6 figure worked out for the minimum monthly contract at 1·47*d.* per copy might be economically more attractive. On the other hand, if the copy print runs were longer and the total daily volume much larger, the offset method (cols. 1 or 2) could show a cost per copy figure lower than that shown in col. 3 even if the col. 3 figures were worked out on the new basis. If larger daily volumes were produced, then items N and K would be spread over the larger volume and consequently the cost per copy would be reduced. Before jumping to any conclusions regarding the "best method" there are two other factors which have not been, and cannot be, taken into consideration in any copy cost comparison.

Mixture of work. The first factor concerns the mixture of reprographic work that has to be done over a given period. For example, not all the originals are likely to consist of dictated material. Some people may say that it is just as quick to type dictated material on a stencil, or on an offset paper master, as it is to type it on a piece of paper. That this is not so, is neither here nor there; the fact remains that some originals cannot be created in their entirety through the typewriter. They may be in a form readily acceptable for immediate electrostatic duplicating without the need to create an intermediate by manual or any other method. The electronically cut stencil method, or either of the electrostatic methods could cope with this situation and either of these would be worth adopting if there were any significant volume of originals which did not need typing. On the other hand, the originals might be too large for the electrostatic machines, in which case a manual stencil cutting process (col. 3) may have to be adopted.

The characteristics of the mixture of work to be done can be determined only by analysing the originals which need either copying or duplicating over a period of at least a month. The headings of any such quantitive analysis would include:

ANALYSIS OF ORIGINALS

Origin	Size	Material Content	Contrast, Colour, etc.
Shorthand notes	A5	Text only	Good
Audio dictation	A4	Text with diagrams	Bad
Pre-prepared	A3	Text with half-tones	Indifferent
	Foolscap	Solid area artwork	Blues, Yellows, etc.

While this analysis is being done characteristics of the output required by the machines to be used can also be recorded. The headings in this case might be: number of copies; need for colours and solid areas; sizes (reduction or enlargement requirement); quality (prestige or managerially adequate): urgency (immediate, same day, as soon as possible). The overall time taken from the initiation of the requirement to the disposal of copies is important because in certain circumstances the most economical method may take too long to produce for operational purposes.

Convenience of method. The second factor is concerned with the convenience of the method as a whole. The through-put of reprographic work might be seriously delayed due to the absence of one or more of the skilled workers needed to produce the stencils or those concerned with the duplication process itself. For maximum flexibility it may be more convenient to use a process which permits anyone to make 1 or 100 copies more or less while they wait, without recourse to a stencil or master and whether the key operator or supervisor happens to be available or not.

The difference in the copy costs between processes gives some indication of the costs of convenience in the particular example given. Another convenience factor arises from the siting of the reprographic equipment. A centrally located machine might induce a queueing factor if it has to serve several departments especially during peak periods which is just when delays cannot be tolerated. A series of departmentally located copier/duplicators could overcome this factor.

The convenience factor can only be assessed by the user in the light of his local circumstances and cannot be reflected in the cost per copy figures. The tabulation and the formulae will, however, provide the user with some indication as to how much he has to pay for the added convenience of the more flexible process. The formulae will reveal a change in the cost per copy as circumstances are varied.

Comparison of cost figures

Any individual check of the calculations arising from the application of the formulae in the circumstances of the example, might provide a different answer. This difference arises from matters of judgment as to what is fair to charge against any particular process. For example, in column 2, the master for the offset machine is made electrostatically. If the master is made as just one job included in a mass of other electrostatic copying, item U (0·8*d.*) would be much less. This would also apply to item W (0·35*d.*). The question arises whether the other copying activities of the electrostatic copier should be used partially to pay for its master making activity for cost comparison purposes.

When comparing the copy costs of the various processes, it should be noted that the typist who manually creates the offset master (col. 1) does not have her master making labour costs partially paid for by her other activities. Only her time in actually creating the masters per day has been charged against the costs of creating copies by this method. Her free time is assumed to have been used on other work.

Like circumstances must be compared with like circumstances where possible. In this case, it has been assumed that the electrostatic copier is not used for any other purpose than to make the masters. The consequent and relatively high cost per copy is an indication to management of the value of the convenience on the one hand, and the need to find other work for the copier so that it shall not be idle most of the time, as in the case of the example chosen.

An alteration in the total copy volume per day would not necessarily have the effect of reducing the copy costs noted in cols. 5 and 6, in equal proportion. This is because the major cost element in col. 5 is the cost of the electrostatically sensitized paper (1·74*d.*) which does not alter as the copy volume increases. The major cost element in col. 6 is the meterage (1·66*d.*) and this reduces with increased usage if each duplicating run becomes longer. It is true to say that the suppliers may often provide quantity discounts for sensitized papers supplied in large quantities on a contractual basis, but this arrangement would only have the effect of reducing the 1·74*d.* figure to a lower figure which would then remain the same irrespective of the number of copies made.

It is unfair to conclude from the figures given in the tabulation that the offset litho process is as expensive generally as the figures seem to show. It is expensive only in the circumstances of the specific example cited. No practising reprographer would, in fact, use the offset litho process exclusively for such a low daily volume of duplicating. Each run would take the offset machine only about 30 seconds to complete and even if 60 seconds were added for make-ready for each run, the whole series of jobs would be completed in about 15 minutes and the plant would be idle for the rest of the day. In these circumstances, it is not surprising that the copy cost by this method appears so high. Under normal efficient work load conditions the offset cost per copy could be as low as 0·75*d.*, all costs included.

Undoubtedly low copy costs can be achieved by centralizing reprographic equipment. By this method of operation machines can be kept loaded to optimum capacity. To ensure this, however, a queueing factor has to be introduced. At the centralized print room originals have to queue for the master making facility and these in turn have to queue for the printing operation.

This costs money in delayed decisions and action. It can also cost money in lost business opportunities. Furthermore, both originals and finished work have to be carried to the print room and returned to the originator. These costs are seldom, if ever, included in the cost per copy. A method of adjusting traditional costs to include these factors is given in the Appendix.

Bearing these factors in mind, the centralized use of reprographic equipment is not necessarily the best and only way of organizing a business visual communication system. Usually a combination of both systems is used.

The examples were deliberately chosen to draw the reader's attention to the need, in general, to use the reprographic process and method most appropriate to the task in hand. The characteristics of the local mixture of reprographic requirements determines the nature, siting and mixture of reprographic products to be used in order to produce optimum efficiency, convenience and economy. These considerations will continue to be relevant while businessmen insist upon creating visual communications in bulk at source, instead of transmitting the information electronically from source to destination and creating lesser quantities of visual communications at each destination for local short-ranged distribution.

CHAPTER X

PROCESS AND MACHINE PROJECTIONS

There is just so much that the design technicians can do to and with established machine design to improve performance. The return on research and development investment tends to diminish in proportions to the previous progress made. Sooner or later, it becomes more economical to wipe the slate clean, so to speak, and start again. In doing so, the designers can incorporate technical advances in materials and technology and, at the same time, anticipate extensions and/or changes in the market's requirements. This stage in the electrostatic reprographic development cycle is about to be reached, it if has not already been reached. Output speeds of a copy a second for line work on the one hand, and the production of acceptable half-tone copy quality on the other, with considerable ease of operation, reliability and versatility, represent the current ceiling of achievement.

The break-through, if it is to come, will have to be concerned with significantly increased speeds of output, significantly reduced costs per copy and the production of true colour copies on both normal and sensitised paper without increasing the cost per copy, or sacrificing the speed and convenience benefits already achieved.

Colour problems

The need to produce at least line colour copies for business and commercial use, has been present for some time. Indeed the absence of a facility to copy in at least two colours has made it necessary for originals to carry credit figures, lost data, reverse trend statistics, etc., in brackets, to show that these figures are different from the mass of normal figures in tabulations and schedules.

A colour copier was demonstrated at the 1968 Business Efficiency Exhibition, by the Minnesota Mining and Manufacturing Co. At that time, exposures were critical and not every type of coloured original was acceptable for copying. The process was relatively slow and the copies relatively expensive. Modern technology being what it is, these limitations are expected to be only temporary. The publicity department of the company have stated categorically that the process is not electrostatic. It is therefore irrelevant to describe the process in this book. It is, however, relevant to consider possible colour copier developments in the electrostatic field.

Before doing so, it should be noted that the major manufacturers in the electrostatic field are divided in their opinions as to whether electrostatic colour printing is likely to be an important future development. In *Business Week* of 2 November, 1968, the chairman of the SCM Corporation is quoted as saying, "We have experimented with the electrostatic colour process and decided that the cost of the equipment, cost per copy and market potential, do not justify the necessary investment". The administrative vice president for marketing of the Addressograph-Multigraph Corporation, in the same article, is quoted as follows: "Who is going to buy it? We have always tried to reach the major market to sell a machine that will do the most for the most people". The marketing manager at Dennison Manufacturing Co's copier division is quoted as follows: "It has a definite application in limited usage in the lithographic business, but it would have little application in general business. The market is not right now, nor in the immediate future. We want a mass market, not a specialized one." The President of the Xerox Corporation, however, told his stockholders in May 1968 that the Company was working on new imaging methods capable of both directly and immediately reproducing in colour, based on a new principle of creating images by photo-electrophoresis.

The photo-electrophoresis process is electrostatic in that the charged toner particles of various colours carried in the dispersant are made to migrate as a result of the action of light energy of various wavelengths. The correct colour can thus be deposited in the correct position on the photo-receptor. The process is in the research and development stage and therefore is likely to be an interesting development for the future.

There is no technical difficulty in producing coloured toners and provided that the demand is high enough, they can, and no doubt will be produced as economically as the black toners are produced at present. Technically, these coloured toners can be dispersed in a liquid, applied by magnetic brush methods of development or cascaded as at present. It is the application of the coloured toner in the right place that will lead to radically altered design configurations and cause an expensive research and development effort.

Line and tone colour

There are three possibilities governing the presentation of coloured originals to the copier or duplicator. The various colours required may be separated at source, as in the case of artwork or each colour can be masked out while one coloured part of the original is copied or duplicated in its original colour and the process repeated for the other colours. The term used to refer to originals of this nature is "line colour". The problem of accurately registering the various colours in juxtaposition has to be solved in both cases. The third case arises when the original contains tonal colours which are merged into one another and cannot be separated by physical means. Optical colour separation techniques have to be applied as part of the copying process. The term "tone colour" is often used to describe this type of original. To handle each of these possibilities, a major re-design of the automated machine is necessary.

For line colour, the copy has to be processed separately for each colour and a multicolour development mechanism has to be built into the machine. Furthermore, the fusing or image fixing method has to be such that the papers on which the colours are deposited can be put through the machine several times without damaging either the paper or the image previously created.

For tone colour not only will the machines have to be re-designed, but the toner characteristics may have to be changed so that when the copies are processed several times, one colour is deposited upon another and made to flow into it at the fusing stage so that the various primary colours are mixed to produce all the other colours. Again, special designs will be necessary, probably incorporating a screen, so that the previously deposited colour, which must remain unfused and unfixed, does not smudge before the next colour is deposited. It is by mixing cyan, magenta and yellow that multicolour prints are produced, so toners in these colours will have to be produced if multicoloured electrostatic printing is to become a commercial reality.

Liquid-development colour methods

There are three major ways of developing a latent electrostatic image. These are the liquid dispersal method, the magnetic brush method, both of which use Electrofax paper, and the cascade method associated with the use of ordinary paper. Colour prints have been produced on Electrofax paper for map making purposes, using the liquid disperal method of development. Although the imaging of normal paper has been associated so far with the cascade method of development, there is no technical reason why magnetic brush, or indeed liquid development methods, should not be used to image normal unsensitized paper by the Xerox method.

The September 1958 issue of the *RCA Review* contained a paper by J. S. Rydz and S. W. Johnson which presented some of the techniques for achieving electro-photographic colour prints and discussed the colorimetric laws concerning the behaviour of Electrofax particles.

Broadly, the liquid development method of producing colour prints depends on the spectral sensitivity of the dyes incorporated in the zinc oxide coating (see page 32). For monochrome reproduction these dyes are added so that the photo-receptor responds to the colours. Theoretically, by allowing the photo-receptor to view the coloured original through the appropriate filter, developing the image after each exposure and by using the appropriate coloured toner, the result will be a direct colour print. After each exposure and development, the photo-receptor has to be recharged. A series of developer troughs, charging and exposure stations form part of the machine's design. Inevitably, such a machine is cumbersome, complicated, slow and only line colour originals can be accepted for processing.

The wide range of sensitivity to the various parts of the spectrum of the photo-receptor suggested a method of separating the colours without the use of filters. With papers incorporating the appropriate dyes in the zinc-oxide coatings, there is approximately, in photographic terms, a two-stop difference between the red and blue spectral colour regions. Thus, by controlling the exposure, separation of the colours can be achieved.

For example, the first exposure to the Electrofax surface with a controlled amount of illumination produces only (say) the blue parts of the image, which can then be developed with blue toner. This development stage renders the developed areas insensitive to further charging. The next step would therefore be to charge the photo-receptor again and reduce the illumination level by a specific amount. This results in the red areas of the photo-receptor becoming sensitive. By developing the photo-receptor in a liquid bath containing red toner, the red parts of the photo-receptor become imaged. The red and green spectral response of the zinc-oxide photo-receptor is approximately the same, but because the deposition of the red and blue toner has rendered these areas of the photo-receptor insensitive to further charging, only the remaining areas have to be re-charged. These can be developed with (say) green toner for a three-coloured copy to be produced using white light for illumination purposes and without the use of filters.

This method is only applicable where the required colours in the final print can be separated at source or are separable by masking while each of the colours of the original are separately processed and where the tonal range is narrow. It is, however, a potentially useful method for colour coding, line diagrams, maps and for the production of coloured copies of advertising layouts.

Dry cascade methods

Theoretically, the dry cascade development method could be used to produce coloured prints from coloured originals given the availability of coloured toners. The colours of a line coloured original could be sequentially filtered out and the original could be sequentially exposed for each colour. Each colour could be developed by means of a number of development stations positioned around a rotating drum each of which could be loaded with a differently coloured developer. Each of the developing stations could be programmed to be brought into operation as each colour is processed.

As each colour was deposited, it could be sequentially transferred and fused on to normal paper. The various colours could thus be built up on the copies by passing them through the colour copier several times. On each occasion the colour filter would have to be changed so that only one colour was developed and fused at a time.

Such a system could cope with line colour, but in order to cope with tone colour development would have to take place in an electrostatic field and each coloured original would not only have to be sequentially colour filtered but screened as well.

Theoretically and technically all this is possible but it would require a very expensive research and development effort. It is not surprising, therefore, that manufacturers are looking for a more simple method by means of which the appropriate colour can be made to migrate to the appropriate place on the copy from a developer reservoir that contains all the colours that are likely to be required. Photo-electrophoresis may provide the answer, but there are as yet many problems to be solved.

For tone colour, both the mechanical development methods, i.e. cascade and magnetic brush, encounter the same difficulty when depositing one colour upon another or in register with another colour, when for example the original is screened. The difficulty arises from the fact that the second sequentially deposited colour must not be allowed to disturb the lay of the first colour deposited. No doubt these problems will be solved in due course if no more simple method is perfected in the meantime.

Output speeds

There are other areas in which developments might take place in the next decade. There is, for example, the ever present demand for increased copy output speed. The ease and convenience of the automated electrostatic copier might be combined with the speed of the offset duplicator. In 1961, the author put forward the idea that short run direct image offset litho masters, once imaged, could be fed directly from the copier into an offset litho duplicator mechanism which would pick up the master, etch it, damp it, ink it and print it. It was to be a flat bed copier incorporating a range of reductions and was provisionally code named the "Xeroduplex". At the time, the concept was too far advanced and nothing was developed.

At the BEMA Exhibition in the USA in 1967, Addressograph Multigraph announced a new machine which was then called the AMCD-1, which operated on similar principles to those outlined above. It used a conventional, zinc-oxide paper based photo-receptor. The image was dry developed by the magnetic brush method. The copy was produced on the specially sensitized paper, which can also be used as a master for the lithographic process. The print consisted of a reverse reading image. The optical arrangements were thus simplified and duplication could be effected by using direct litho methods and these also simplified the mechanism.

The conventional offset blanket mechanism could be eliminated. Having been imaged, the master was automatically conveyed and fixed to a direct litho impression cylinder. A combined ink and fountain system was used in the form of a removable cartridge. Operational control was effected by the use of press buttons and machine minding was reduced to a minimum.

Once the original had been positioned and the programme set to produce a given number of copies, the whole process was automatic and the original was ejected from the system as soon as the duplicating run started. This enabled another original to be positioned and a fresh programme initiated before the previous print run was completed. The machine stored the new programme until the previous duplicating run was completed and then automatically proceeded with the next programme.

The machine has not yet appeared on the European market although it was originally scheduled for the American market in 1968. When it does appear, it will represent a break-through in the electrostatic zinc-oxide duplicating field. It is possible that machines very much like the one described may become commercially available in Europe as soon as the field trials have proved successful.

An alternative reprographic system combining an electrostatic copier with a fast and fully automated offset litho printer may become competitive in the 1970s with the all-electrostatic duplicators. It is called the Addressograph-Multigraph Total Copy System. The elements of the installation comprise an A-M 2000 electrostatic copier, a 122 Converter unit and a Multilith ASC (automated sequence control). This new machine is a development of the Multilith 1250 unit. The copy, created on specially sensitized paper, is passed through the converter unit and straight on to the offset printer which is designed to run continuously without any operator attention other than dropping the master into the selection tray, and selecting the quantity of copies required.

Canon NP system

Another electrostatic method of copying and duplicating may have a competitive, if not revolutionary affect on the electrostatic market in the early 1970s. This is the new Canon NP system which possibly has been developed from the work done by Kallmann, Rennert and Sidran under the general heading of Persistent Internal Polarization (PIP) published in 1960 and anticipated by Carlson as a possible electrostatic method of document reproduction in his US patent issued in 1939. A fuller account of the way PIP works is given by R. M. Schaffert, MA, PhD in his book *Electrophotography*.

The persistent internal polarization of a photo-receptor requires:

1. A multi-layer photo-receptor consisting of a semi-conductor, or a combination of semi-conductors, a conductive base on which these are bonded with a suitable resin binder, and a transparent, insulating covering.
2. A machine component configuration that allows AC and DC charges of alternative polarities to be applied before and during

exposure, and a low frequency radiation (say infra-red) to be used to normalize the photo-receptor after the copy has been
3. made so that the photo-receptor can be used repeatedly.
A method similar to those described in previous chapters for making the electrostatic image visible and for transferring this to a conveniently usable media.

The science of the process is too complicated to be described in detail in a general study such as this, because it involves an understanding of solid state physics. Briefly, what happens is that depending on the technical nature and relative thickness of both the semi-conductor and its insulating covering and their electrical compatibility, the electric charges representing the images are trapped for long enough to permit them to attract toner and for this to be transferred to a convenient, unsensitized copy media.

Elements of PIP method

Assuming that the Canon NP process is a development of PIP methods, the probable elements of the process can be summarized as follows:

1. *Photo-receptor.* This would consist of a conductive base of aluminium or zinc. A suitable resin binder may be used to bond a semi-conductor layer to the base. The semi-conductor might consist of cadmium sulphide (CdS), cadmium selenide (CdSe), a mixture of these, or, indeed, crystalline selenium which is unsuitable for xerography. The semi-conductor layer would probably be coated on the base by vacuum coating techniques, or by spraying. The photo-receptor in turn would be coated with a transparent insultating material such as Mylar.
2. *Pre-charging stage.* This might or might not be necessary, depending on the type of semi-conductor used. A direct current corona discharge mechanism would probably be used and its polarity would depend upon the type of semi-conductor or photo-receptor chosen.
3. *Charging and simultaneous exposure stage.* The lighting of the original would be more or less as described on page 57 with the significant difference that an alternating current corona discharge field would be positioned across the optical path.

4. *Illumination of photo-receptor.* An ordinary tungsten filament lamp would probably be used to photo-flood the whole of the photo-receptor after the stage described above had taken place.
5. *Development.* The toner powder could be applied to the photo-receptor by the cascade, magnetic or fur brush methods already described. As the polarization is persistent, however, the development stage could take place in ordinary light.
6. *Transfer.* The image could be transferred on to practically any type of smooth, flexible media. A transfer charge has to be applied and this could be done by either a corona discharge arrangement or by the use of a charged roller.
7. *Fusing.* The conventional method of fusing the powder image on to the transfer media (see page 69) is likely to be used.
8. *Pre-cleaning.* The usual method of loosening the residual image after transfer would probably be adopted, that is to subject the residual image to a suitably polarized corona discharge.
9. *Physical cleaning.* The residual image, having been loosened electrically at the previous stage, would probably be removed by a rotating toner cleaning brush.
10. *Electrical neutralization.* This is most likely to be effected by subjecting the photo-receptor to a low frequency radiation such as that generated by an infra-red lamp.

The explanation of the process given above is only accurate so far as it can go in a general work such as this. More precise technical details of the process can be obtained by a study of the various patents that the Canon Camera Company has lodged.

Competitive processes

In March 1961, the *New Scientist* reported on the possibility of documents being copied by "selective distillation" methods. The invention is attributed to Mr. Abram Games. Briefly, the process requires the original to be oiled and brought into contact with unsensitized paper. Both are then momentarily heated by an infra-red lamp. The heat is absorbed by the black areas of the original and reflected from the surrounding white areas. The heated oil on the black areas of the original becomes distilled on the cooler copy paper with the result that the copy paper carries a latent image of the original in the form of an oily transparent fluid. The original is returned to the user and the copy paper passes through a powder thermosetting ink trough so that the particles of ink adhere to the copy paper, or any other media that is being processed, only where there is an oily image. The copy can then be subjected to heat, which fuses the image into the copy paper surface.

A development of this process was reported in *Chemistry and Engineering News* in January 1968, and called Tempo. The advantages of the new process arise as a result of using an absorption of reduced layer formation process developed by Mr. Games in 1965. A special liquid is used to coat the original and the process depends on the difference in the absorption rate of the liquid from one colour to another.

The result of this differential absorption is that liquid is left lying on the less absorbent areas, and a liquid pattern of the entire image is produced which can then be transferred by contact on to a sheet of unsensitized copy paper and the toner particles picked up as described above.

The original process was limited to images which would become heated by the infra-red radiation and so was blind to coloured inks and those inks which did not contain carbon black or other metallic compounds. The improved process uses differential absorption techniques and can register all colours. By reason of the fact that the oiled original comes into direct contact with the copy paper face to face, the image when developed is a reverse reading image. Provided that the copy paper is translucent enough, the copy can be read as a right reading image from its reverse side. No doubt this limitation will be overcome in due course if the process is to become a viable commercial proposition.

The *Financial Times* of 16 January, 1969 carried an interesting report concerning a low cost photo copying machine manufactured and marketed by Leed, a subsidiary company set up by Lines Brothers. The machine is called the Copymate and is being produced under licence from the Transogram Company in the US. This company developed the equipment in association with the Minnesota Mining and Manufacturing Co.

The original is placed between two sheets of specially sensitized paper and exposed to light. The two sheets of paper are then placed together on a second panel where the heat process develops a positive image and so produces the final print. The machine is now marketed at about £15 and the cost per copy is quoted at one shilling. It is an interesting development which may be worth watching for its competitive impact on the electrostatic copying and duplicating market, especially if the cost per copy can be reduced.

Another interesting development which may affect the amount of hard copy created by electrostatic machines, is the possibility of transmitting original material direct to recipients by means of the telephone communication copier. The system is not electrostatic and the implications are likely to be that less hard copy will be created at source and physically transmitted in its original form to the recipients.

The sender and the receiver would each have a machine attached to his internal or external telephone. In order to transmit copy, the sender would merely have to make a normal telephone call, ask the recipient to place his telephone handset on the Telecopier and wait. The sender would also do the same and then insert the visual communication to be transmitted into the sender's Telecopier. In about six minutes a facsimile copy of the original document remotely transmitted, would be reproduced by the receiver's machine. A schematic diagram of the arrangement is shown in Fig. 25.

Long-distance xerography

There is an electrostatic version of the Telecopier system known as long distance xerography. It is operational in the USA where there are, at present, more suitable broad band electronic links than there are in Britain. As the demand for faster transmission of information crystallizes and as more national and suitable transmission links become available, this process will become more commonplace.

The apparatus consists of a scanner at the transmission end of the link and a printer at the reception end (see Fig. 26).

Once a visual communication has been converted into electronic signals and transmitted by either line or microwave link, the emerging electronic signals can be converted into the language of man in many ways. They can be in the form of light signals scanning a photo-receptor or they can be applied to zinc oxide coated papers or plain papers or by the Videograph or Videojet methods. The equipment is interesting in itself but it is the implications of the possible adoption of these systems that could have a profound effect upon the type and use of electrostatic copiers in offices, and the effect on the training of reprographers.

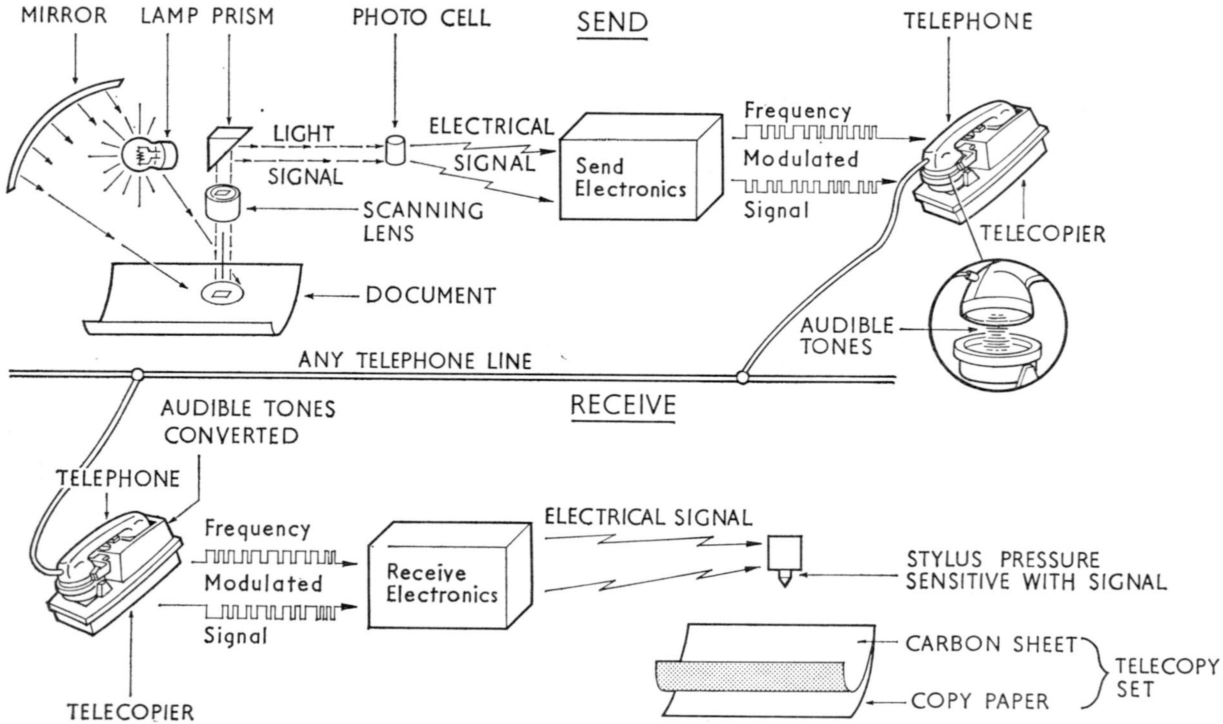

Fig. 25. Schematic diagram of the Telecopier electronic information transmission system.

Storage and retrieval methods

After it has done its job of communicating data, information, diagrams and pictures to the mind of man, the ultimage destination of the visual communication is the filing cabinet. It follows, therefore, that more convenient, quicker and totally different methods of information storage and retrieval would have a profound effect upon the way visual communications are handled, created, copied and duplicated. Such a new method has been invented and although it is at present being applied to the filing and retrieval of engineering drawings, there is no reason why the system could not ultimately be

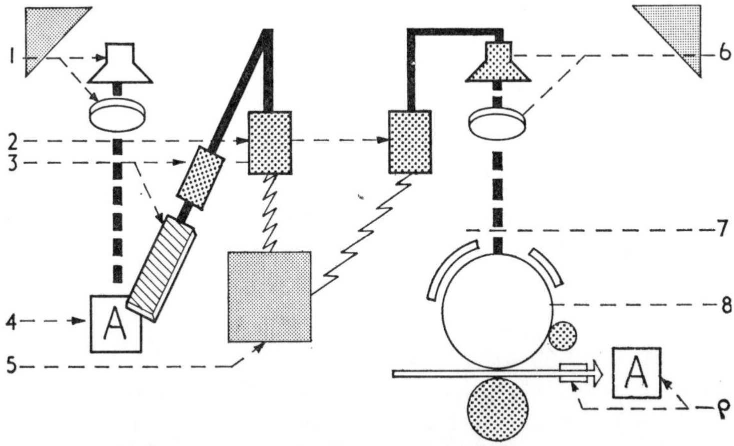

Fig. 26. Schematic diagram of the Xerox LDX electronic Information Transmission System: 1, cathode ray tube and lens scanning the original (4). 2, scanner and printer electronics. 3, light guide and photo multiplier. 5, broad band transmission channel. 6, receiver cathode ray tube and lens. 7, development and charging section of selenium drum (8). 9, image transferred and fused to paper to form output reproduction.

applied, perhaps in modified form, to the storage and retrieval of business documents.

In November 1967, there appeared a report in *Reprographics* of a joint project to develop video techniques for the high speed filing and retrieval of engineering drawings and similar documents. Sandia Laboratory, Albuquerque N.M. is reported to have awarded contracts to Ampex Corp., of Redwood City, California for the development of a Videofile document filing and retrieval system.

The system uses television techniques to reduce conventional files of documents to compact magnetic recordings on videotape. A

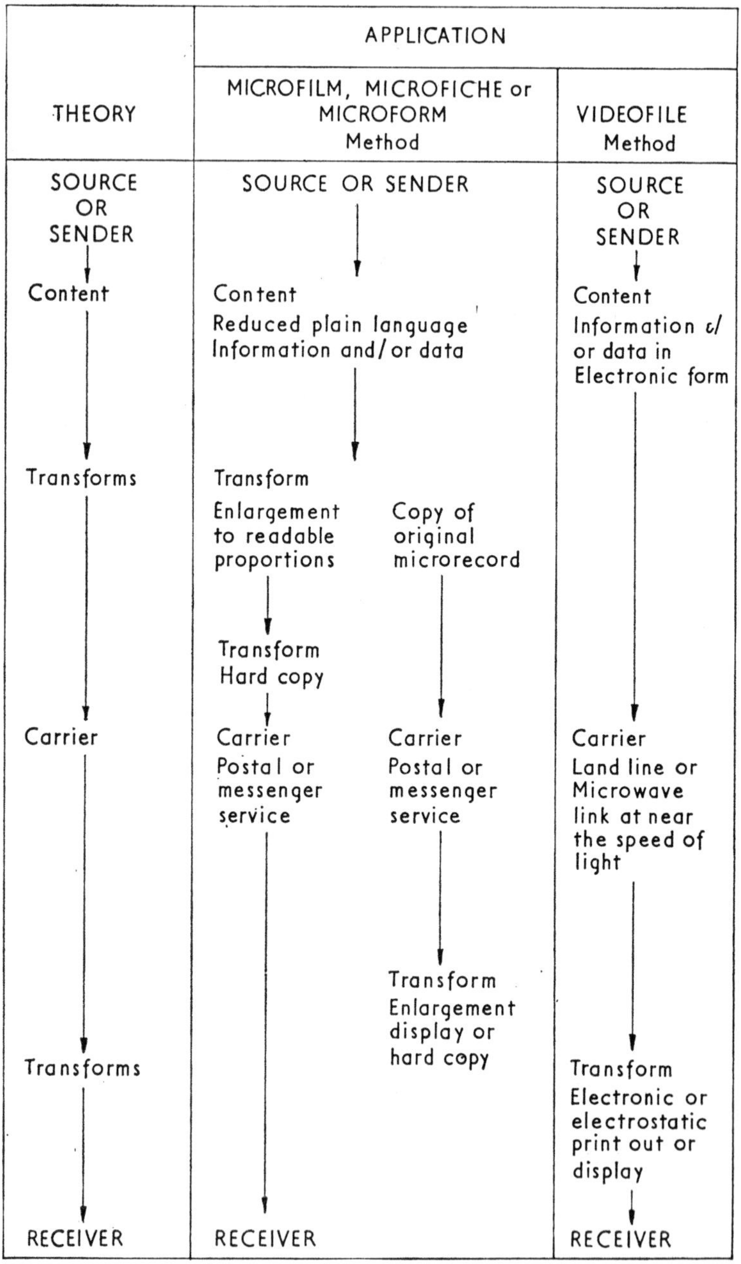

Fig. 27. The anatomy of the communications link.

single 14 in. reel of tape is able to store 250,000 pages. The stored information can be located, viewed on television screens or reproduced as facsimile copies in seconds. The properties of magnetic tape permit documents to be added to existing tape files or they may be erased, relocated or replaced electronically within that file. Interesting though the technical equipment may be, the implications of the impact that the adoption of such a system would have on management of visual communications are of more concern to reprographers.

The Videofile system is likely to be more useful for data and information storage and communication purposes than microfilm, microfiche or microform methods because of the nature of the Anatomy of the Communication Link (Fig. 27).

All communications whatever their nature follow the pattern set by the theory. There has to be a content, otherwise there would be no message. The content has to be transformed into a form convenient for the carrier. If it is a letter, then it has to be written, have an envelope, an address intelligible to the carrier, a stamp on it and it has to conform physically to the postal regulations. In this particular case, the second transform is unnecessary because the message was originally created in a form acceptable to both the carrier and the receiver at source.

It is obvious that by reducing the number of transforms that are necessary and by speeding up the mechanics of each transformation, the speed with which the communication can be made is increased.

Fig. 27 shows that many transforms are necessary in the microfilm or microfiche methods before the message arrives at its destination in usable form. Furthermore, the message has to be carried by relatively slow methods. The advantage of the Videofile method can be seen at a glance. The content, in this case, is stored in electronic form, capable of direct and very fast transmission which needs to be transformed only once into the language of man at destination. The mechanics for storing the information in the first place are no more complicated for any one method than for the others. At present the microfilm system is, however, the more economic.

Over the next ten or fifteen years microfilm and microfiche storage and retrieval systems will develop and become much more sophisticated and perhaps cheaper to operate than they are now. By the 1990's however, electronic storage, retrieval and transmission systems can also be expected to develop and become more economical. It is anybody's guess as to which of the methods will ultimately prevail. Business and commerce, however, can be expected to go into the 21st century with much more efficient methods of storing, retrieving and communicating information and data than are commonly used at present. If and when visual communications are created at destination, rather than at source, the copiers and duplicators in use then will be very different from those that we know now. The reprographic education and training of businessmen will also have to develop if the emerging technologies are to play an important part in the communication systems likely to be in use towards A.D. 2000.

APPENDIX

THE COST OF INCONVENIENCE

In this age of instant coffee and instant mashed potato, anything less than instant duplication is a costly business inconvenience. For some obscure reason the established traditional methods used are taken as the criteria against which the modern methods should be judged. That the traditional methods were the "best buy", indeed the only "buy" at the time of their debut is no reason why they should continue to carry the accolade of perfection in the changed circumstances of today.

The difficulties in arriving at a realistic cost comparison between the traditional and modern methods of duplicating arise because the true nature of the function is not understood. The duplication of visual communications, without which business and commerce would come to a halt, is usually considered to be a means to an end in itself, and is costed as such and therefore costed out of its context, whereas the duplication of documents should be considered as part of a service which links the originator and the recipient. The duplicating function as such is only part of the whole communication link.

It follows therefore that to say that the cost of duplicating is the cost of communicating is dangerously misleading because there are many other inconvenient cost factors which are not usually taken into consideration. It also follows that the cost of the inconvenience of the traditional methods should be taken into consideration when cost per copy figures are used to compare the economics of the traditional and modern methods of duplicating. Before proceeding to show how surprisingly easy it is to cost the inconvenience associated with the production of visual communications, it is necessary to examine the environment in which they are created if only to establish a common ground for a meaningful discussion of this important subject.

The originators of visual communications need to use two reprographic functions—the copying and the duplicating functions. Nowadays, the secretary who creates or assembles material to be duplicated no longer has a duplicator for her exclusive use, but she often does have a copier. She has however the use of a duplicator, or duplicating services remotely located somewhere else. It is located in a typing pool or print room where it gets more intensive utilization in order to secure a greater return on the investment in the asset.

At the print room the document has to take its place in the queue and a master intermediate has to be made prior to printing. The master then incurs queueing time waiting for the duplicating machine be it a stencil or an offset litho unit, to become available. Further time is taken while the machine is made ready and the prints made. If the original is a multi-page document, still more time is taken up while the copies are manually collated and bound or stapled together. The original and the copies are then carried back to the originator for despatch.

The costs of these inconveniences, the lowering of the individual productivity and the cost of the delays involved are seldom taken into account when the costs of the traditional methods of duplicating are worked out. It is erroneously assumed that the same steps are necessary when modern methods are used.

The modern method of duplicating permits the secretary to have in one machine both a copier and a duplicator. The machine is press-button operated and can be located within a few paces of the desk. Furthermore, it does not need an intermediate and once programmed and switched on, it operates by itself and switches itself off, so releasing the secretary for other productive work. In fact, by using these modern machines she can be said to be doing two jobs at once.

There are as many different ways of creating any particular communication as there are trees in a wood. Those responsible for producing cost comparison figures between the traditional and modern methods of duplication, examine metaphorically each tree and its immediate environment in the greatest possible detail. From their detailed examination the business backwoodsmen and their army of foresters pass on their conclusions to management with the advice that this or that is the "best buy" as a matter of policy. It is as illogical to deduce the general conclusion from an isolated example in costing out a reprographic task, as it is to deduce the character of a nation from the observation of one individual. The man at the top of any particular functional tree knows better. He knows that the welter of differing cost per copy figures produced from various sources arises from the fact that too many variables vary too frequently and that there are many hidden costs which cannot be accurately calculated, and are therefore ignored.

Can the hidden costs of using the traditional methods be uncovered so that true cost comparisons can be made? There *is* a way, provided that it is agreed that "time is money" and, for the purpose of comparison only, that an original once created attracts a proportion of the overheads just as does an employee, or a piece of machinery when he, she or it operates within the business environment. Furthermore, there is a way of "price tagging" time.

It is a fact that there is a limited number of business minutes in a year in which to translate a potential loss into a profit. It is also a fact that overheads mount up with time and the annual overheads apportionable to the company as a whole or any particular location, can be known. An operating cost per minute can be calculated by dividing the annual overhead by the number of business minutes. This expense continues whether the time is productively used or not.

Assuming, for example, that the annual overhead for any particular function is £10,400 pa the cost per minute can be worked out at 20*d.* With these basic facts, which could vary from company to company, it is possible to "price tag" time and to price the inconvenience of one method against another. This figure can then be equated to the cost per copy and shown as an additive constant by which the cost per copy figure calculated by the traditional methods should be adjusted. If, for example, 60 copies of a ten-page document are to be duplicated on both sides of the paper to reduce transit costs and the amount of paper used, the traditional and modern methods can be compared to discover the ratio linking the two.

By modern methods the time taken to do this job is as follows:—

Time taken from executive desk to electrostatic printer in secretary's office	1 minute
Printing time one side	5 minutes
Printing time second side	5 minutes
Changeover of paper	1 minute
Changeover of originals	1 minute
Collation is automatic	
Stapling	3 minutes
Time from stapling on secretary's desk to out tray	1 minute
TOTAL	17 minutes

By the traditional methods, say offset litho, the time taken for the job is conservatively estimated as follows:—

Time taken from executive desk to print room	5 minutes
Queueing time for master making....................	5 minutes
Master making—10 masters..............................	15 minutes
Queueing time for printing............................	5 minutes
Make ready time (each master one minute)..........	10 minutes
Total printing time..	4 minutes
Manual collation and stapling..........................	20 minutes
TOTAL	64 minutes

Using the overhead figure of 20*d.* per minute, the modern method incurs an inconvenience factor of $17 \times 20 =$ £1. 8. 4*d.* which when spread over the 600 copies works out at ·56*d.* per copy. The traditional method incurs an inconvenience factor of $64 \times 20d. =$ £5 6. 8*d.* which when spread over 600 copies works out at 2·1*d.* per copy.

The cost of inconvenience of the traditional methods as compared with the modern methods is in the ratio of approximately 4 to 1. If the actual cost per copy difference is required then the two figures calculated above should be added to the normally calculated figures for the two methods used in isolation. The actual figures given in the tabulation above would vary between companies but could be easily determined by work study techniques. The figures given are only to illustrate a method by which the costs of the traditional methods can be objectively compared with the costs of the modern methods. To overcome the objection that these are two isolated examples, the respective data could be assembled over a period of a month.

In straining for maximum utilization of traditional reprographic assets, those who advise this course automatically ensure that the assets are centralized. They are therefore removed from where the services are needed and so incur carrying and delay costs. The documents must queue up for attention and thus incur the costs of operationally inefficient communications, in these circumstances.

The modern methods are geared to modern circumstances, whereas the traditional methods are not, therefore it is the traditional methods that should be carefully examined to see if they measure up to modern circumstances. The modern methods make duplicating as instant as modern technology allows. The fetch, carry, queue and return delivery costs which add to the prime costs sometimes by a factor of 4, can be eliminated if modern copier/duplicators are operated at the point of need, i.e. source for which they are designed.

Office copiers and duplicators are the tools of efficient management along with typewriters, telephones, desks, calculators and computers. A machine tool is valued for its instant availability and its contribution to the overall efficiency of the production system. So should copier/duplicators be considered vis-à-vis the visual communication system. After all, one does not deny oneself the use of a machine tool because it can only be used an hour or so per day, neither does one deny oneself the use of a telephone because one cannot talk into it incessantly.

In business we should get our priorities right. The overall efficiency of an integrated visual communications system is more important than any one element of it. By installing electrostatic push button operated copier/duplicators, which make and collate copies unattended at the point of origin and therefore need no labour, copy costs can be reduced and the visual communication efficiency improved. Once the system as a whole is working efficiently and in harmony with corporate objectives, then the efficiency of each component of the system can be examined, improved and made more economical.

Pressureless printing allows things to be printed on without being touched. It raises possibilities of printing on soft fruit, shaving cream and even the yolk of an egg. It is an electrostatic process and the equipment consists basically of a fine mesh metal screen bearing a stencil and a conductive backing plate parallel to it a short distance away. A very fine toner powder is applied to the screen where it acquires a positive charge, either by the tribo-electric effect of the friction by the ink applicator on the screen or by charge transfer from the screen, or both. The positively charged particles are attracted through the screen and across the gap towards the negatively charged backing plate.

The article to be printed is interposed between the screen and the backing plate so that a powder image, conforming exactly to the pattern on the stencil, is deposited on it, in much the same way as the image on a Xerox plate or drum (see Fig.)28.

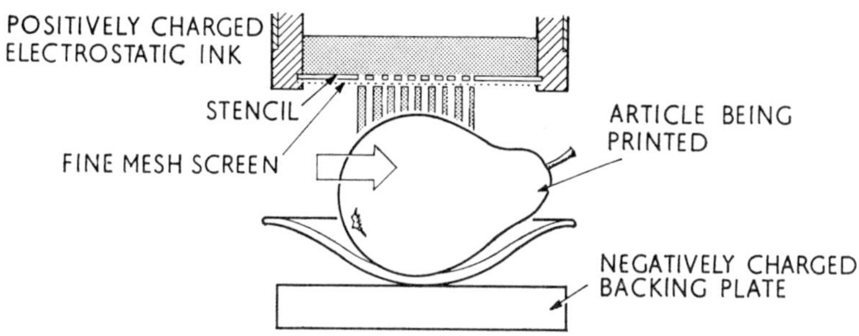

Fig. 28. Creating images on solid objects electrostatically by pressureless printing methods.

The potential is applied to the rollers by means of a brush electrical contact, and the image is fused by passing between infra-red heating lamps.

Preparation of the stencil screen is comparatively simple. In one method the screen is coated with a bichromated gelatine solution and allowed to dry in the dark. This light sensitive gelatine is exposed to ultra-violet light through a photographic film positive, whereupon the exposed (non-image) areas are hardened. The unexposed (image) areas are washed out by soaking the screen in hot water, the screen then being immersed in a hardening solution to give the gelatine more durability.

Good results have been obtained using screens of 133 lines to the inch, but up to 250-line screens have been successfully used. The ultimate fineness of the screen is limited by the size of the ink particles, which are composed of resin and colouring matter.

Each particle of the ink must have the same composition and electrical behaviour as all the others. The size is usually between one and five microns.

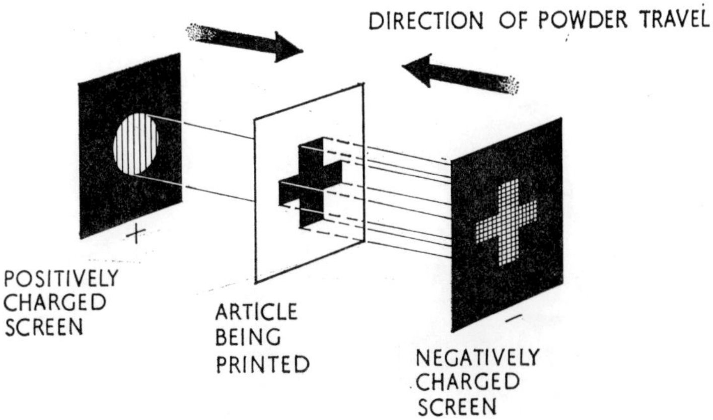

Fig. 29. The simultaneous printing on both sides of paper by pressureless printing methods.

The gap between the screen and the backing plate is variable, but as the gap increases so does the voltage required. Whereas a gap of 0·02 in. needs a voltage of only 300 volts DC, one of 0·0625 in. requires about 2,000 volts. Where the article to be printed has sufficient conductivity, it can also function as a backing plate, and fruit, hot glass and paper can be made to function in this way.

Colour printing is possible. There is no trapping of the ink and several layers of colour may be applied and subsequently set by a single fusing operation.

The obvious advantage of pressureless printing is the ability to print on a wide range of materials. In the USA, a machine produced by Unimark Corporation of San Ramon, California, is printing avocado pears with the name of the grower in edible ink, fixed by a confectioner's spray. In Idaho potatoes are being marked the same way, as are apples in the state of Washington—at 2,200 a minute.

The Owens Illinois Glass Company is developing equipment for printing on glass and ceramic containers. Sun Chemical Corporation, of New York, which also produces the powdered inks used in the process, is collaborating with Fritz Buser Limited, a Swiss textile machinery manufacturer, in printing fabric by pressureless printing, as a faster and cheaper alternative to silk screen printing.

The image is subsequently fixed by heat or solvent vapour. The surface configuration of the article being printed is immaterial because the ink particles travel in a straight line.

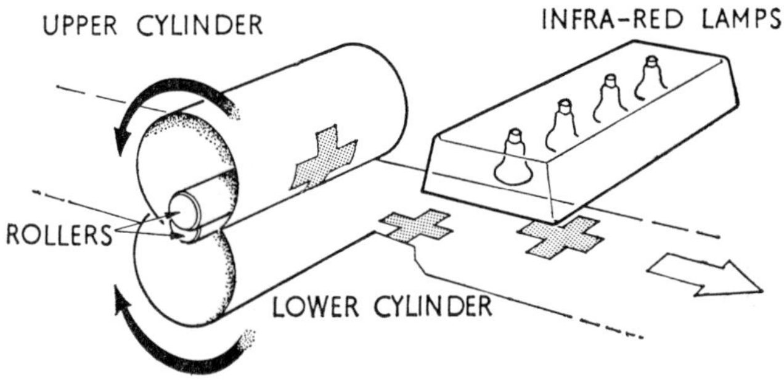

Fig. 30. A rotary method of printing on both sides of paper by the pressureless printing method.

Almost any medium is suitable for printing by this means except one that could interfere with the electrical field, such as a plastic sheet carrying a static charge.

If instead of a backing plate another screen is used then simultaneous printing on both sides of the paper is possible (see Fig. 29).

The principle can be converted to a rotary action by having two screens clamped around the upper and lower perforated cylinders with a toner applicator inside each cylinder. By synchronizing the toner discharge with the rotation speed of the cylinder and the rate of travel of the paper or other material, a high rate of printing can be achieved.

The basic patent rights are held by Electrostatic Printing Corporation of San Francisco, which has granted patents for development in specific areas to other companies. The process was invented by

Clyde Childress and Louis Kabell and the feasibility studies were done by the Stanford Research Institute in Menlo Park, California. The technique is also being developed by Dainippon, a Tokyo ink and chemicals company.

Toner powders are at present dearer than normal printing inks but it is claimed that less is needed. On the operating side everything is far less critical. Many of the complex problems of even ink transfer, close pressure tolerances, drying, and so on, are eliminated. As for those messy washups—five minutes with a vacuum cleaner and it's done.

Reprinted from *Repro*, August, 1966

DECIMAL PENCE CONVERSION

Although the cost figures in this book are given to one or two places of decimals, they are in fact in the old currency. The following table gives the decimal currency equivalents.

Old Pence	*New Pence*	*Old Pence*	*New Pence*
0.01	0.0042	0.50	0.2083
0.02	0.0083	0.60	0.2500
0.03	0.0125	0.70	0.2916
0.04	0.0166	0.80	0.3333
0.05	0.0208	0.90	0.3750
0.06	0.0250	1.00	0.4166
0.07	0.0291	1.50	0.6249
0.08	0.0333	1.75	0.7280
0.09	0.0375	2.00	0.9571
0.10	0.0416	3.00	1.2490
0.20	0.0833	4.00	1.6664
0.30	0.1250	5.00	2.0830
0.40	0.1666		

GLOSSARY OF REPROGRAPHIC TERMS

ACTINIC LIGHT – Chemically active light mostly of violet and blue rays effective in producing photochemical reactions. Obtained from arc lamps, mercury vapour lamps, photo-flood bulbs etc.

ALUMINIUM PLATE – Flexible printing plate used in the offset litho process to which the image is applied mechanically, photographically or xerographically.

APERTURE – An opening (often variable) controlling the passage of light through a lens. Also referred to as the "stop", it is usually formed by an iris diaphragm which can be set to various openings marked by *f*-numbers. The *f*-number is obtained by dividing the focal length of the lens by the effective diameter of the aperture.

APERTURE CARD – A card, usually of a size suitable for use in a Data Processing system, with a rectangular hole or holes specifically designed to hold a frame or frames of microfilm.

ATOM – The smallest particle of an element which can take part in a chemical reaction.

AZO DYE – Dye formed by reaction of diazonium salts and tertiary amines or phenols couplers, as in the diazo process.

BASE MATERIAL – The paper, cloth, film, metal or other supporting medium on which a reproduction is made.

BLANKET – A resilient rubberized surface onto which an image from a master or intermediate is transferred after inking in the offset litho printing process.

BLEED-OFF – An image is said to bleed-off when it runs off the edge of the paper so that there is no white margin at the paper edge.

BLOCKING OUT – The action of deleting or masking areas on a master or an original so that they do not reproduce.

CASCADE – The action of tumbling a powder or a mixture of granules over an object which is to be treated by this method.

CHARGE – The action of causing an object to possess an excess or a deficiency of negative electrons.

CHARGING BAR – The colloquial name given to a corona discharge unit.

COATING – A chemical solution, suspension or emulsion applied to a base material for processing or protective purposes.

COLD FLOW – The partial fusing together of the individual particles of a thermosetting plastic powder due to the mass of the powder being subjected to a high temperature.

COLLATE – To assemble separate sheets together in correct sequence.

COMPUTER TABULATION – The output of data from a computer that has been printed in columnar form.

CONTACT PRINTING – To print by direct contact between a master or an intermediate and the copy support medium giving a same size print.

CONTINUOUS TONE – The gradual and continuous shading of tones (full tone) on an original or copy from white to black or from any two shades of grey or other colour(s). See also half-tone.

CONTRAST – The brightness difference between the light and dark tones of an original or copy.

COPY – A document or collection of written matter for processing or reproduction; an exact reproduction of the original document; to make one or more exact reproductions of the original at one operation without using an intermediate or a special master.

CORONA DISCHARGE – The action of spraying electrical charges (ions).

COROTRON – An electrical device without a screen for creating a corona discharge.

CREDIT TRANSFER ADVICE – Documents by means of which monies may be paid from one banking account to another on the instruction of a bank.

CURL – Tendency to coil or roll up at the edges in paper usually following the grain direction. See grain.

DAMPING SOLUTION – A solution of water, gum arabic and various types of etches used for wetting lithographic printing plates.

DENSITY – The degree of resistance to the passage of light or other radiation of an image or material. It can also be described as the light absorbing power of an image or material.

DEVELOPMENT – The action of rendering a latent image visible.

DIAZO (DYELINE) – A process in which the print results from the effect of light on diazonium sensitised material which is exposed after illumination to ammonia gas or liquid azo dyes.

DIAZO BOND – A paper visually opaque but translucent to ultra violet rays.

DIFFUSION TRANSFER – A process in which the image is transferred by diffusion of non-exposed silver salts from a master to specially coated but non-sensitive material creates a positive image.

DIRECT-IMAGED MASTER – A duplicating or printing master upon which the image has been directly typed, written or drawn.

DIRECT POSITIVE – A chemical process which produces a positive image from a positive original without the intermediate use of a photographic negative.

DISPLAY TYPE – Comparatively large type used for headings, title pages, headlines and posters.

DOCTOR BLADE – Part of the machinery used in the gravure printing press which removes excess ink from the printing surface before the impression is made.

DUAL SPECTRUM – A two-stage process in which a master is first made by the action of ultra violet light and subsequent copies are produced from the master by infra-red light.

DUPLICATOR – A device for making multiple copies of a document usually with the aid of a specially created master.

EDGE EFFECT – A term used to describe the appearance of that part of an electrostatic copy where the outside edge of a character or image is darker than the rest of the character or image.

ELECTROFAX – The name used by The Radio Corporation of America to describe the electrostatic process which uses paper coated with zinc oxide and other chemicals upon which to develop the images and so create copies.

ELECTROGRAPHY – A process used for producing copies on insulating surfaces or by placing the electrostatic charges where, and in the form they are required which are then developed by precipitation on the surface of electrically charged toner particles.

ELECTRON – An electrically charged particle, or element of charge having inertia, which is the most numerous constituent of matter. Although normally existing within the system known as the atom, electrons may be isolated from matter (as in cathode rays). The electronic charge e is $-4 \cdot 774 \times 10^{-10}$ ESU, its mass when at rest m_6 is $9 \cdot 042 \times 10^{-28}$ gm.

ELECTROPHOTOGRAPHY – A process used for producing copies on or from an electrically charged photosensitised semi-conducting surface by using light to remove the charges from where they are not required and by leaving the charges where they are required of latent images which are subsequently developed by precipitation on the surface of electrically charged toner particles.

EM – The square of the body of any size of type; a printer's unit of measurement.

ENLARGEMENT RATIO – An expression of the number of times a copy or image is larger in linear dimensions than the image from which it was made, e.g. 5:1 or 5X enlargement.

EXPOSURE – The action of submitting any sensitised or sensitive surface to radiation, normally light or heat, upon which this will act to form an image or a latent image.

FOCAL LENGTH – The distance from a lens to the plane in which it forms a sharp image of a distant subject. The shorter the focal elngth, the greater the power of the lens to bend light rays and thus the wider its angle of view.

FINISHING – The reprographic activity which takes place after the copies have been produced but before they are ready for use. The activities include cutting, collating, stapling, eyeletting and binding.

FIRST GENERATION COPY – The copy or copies created direct from the original, the duplicating intermediate or the original photographic negative as distinct from copies made from the copies originally produced.

FOUNT (OR FONT) – An assortment of characters containing all letters (upper case, lower case, or both) and sometimes figures, punctuation marks, mathematical and other signs.

FORMAT – The size, style, shape and general appearance of any printed material, but specifically the size and shape of a page.

FOUR-UP – A term to describe the reprographic processing of four originals at one time on one sheet of paper.

GELATINE TRANSFER – A process in which photographically produced dye images are physically transferred from a light sensitive master to special copy paper.

GRAIN – Paper fibres which lie chiefly in one way, giving grain direction.

GRAVURE – A commercial method of printing obtained from minute ink filled cells. The deeper the cell the greater the weight of ink applied to the paper.

GUM ARABIC – A fine yellow or white powder soluble in water: sp gr 1·355. It is obtained from certain varieties of acacia, the world's main supply coming from the Sudan and Senegal. It is also called acacia gum, or Senegal gum. It is used reprographically to coat offset litho printing plates to prevent the marking and oxidisation of the surface both before and after printing.

HALF-TONE – The representation of tonal graduation by an image composed of dots of varied sizes, the centres of which are equidistant.

HARD COPY – An eye-legible document or copy, usually on paper.

HECTOGRAPH – see spirit duplicating.

HECTOGRAPH CARBON SHEET – A specially prepared sheet of carbon paper containing aniline dyes used to create a reverse reading image on a master sheet used in the spirit process of hectographic duplicating.

IMAGE – A likeness of the subject matter.

INFRA-RED LIGHT – Light rays having a longer wavelength than visible light rays.

INTAGLIO – A printing process in which a master is used into which the image has been engraved or otherwise recessed.

INTERMEDIATE – The special medium needed by some copying, printing and duplicating processes that has to be created before copying, printing or duplicating can take place. The type of medium varies with the type of machine with which it is to be used.

IONIZE – To cause substances to have an excess or a deficiency of electrical charges.

IRIS – A continuously variable hole forming an adjustable stop for a lens usually integral with the lens mounting.

JUSTIFICATION – Even spacing of typeset matter within a specified line length.

LATENT IMAGE – The image existing in or on a photo-receptor after exposure but before development.

LATERALLY REVERSED – A mirror image which may be either a positive or a negative copy of the original (see also right reading).

LEAD – To make or widen spaces between lines of type.

LEADED – Lines of type separated by additional space.

LENS – An arrangement of curved glass surfaces which collect the light reflected from or projected through an original document or object which is to be reproduced, and distributes the light on to a photo-receptor. The light is focused by positioning the lens in relation to the object and the photo-receptor surface.

LETTERPRESS – A printing process which the printed image is obtained by the transfer of ink from raised type, or from half–tone or line blocks.

LINE ORIGINAL – An original with no tonal graduation, either real (see continuous tone) or simulated (see half tone).

LITHO – Short for lithography or lithographic according to context.

LITHOGRAPHIC RIBBON – A special ribbon used in typewriters and other similar machines for creating by direct methods images on offset lithographic masters.

LITHOGRAPHY – Originally the art of printing from stone, but now applied to printing processes depending on the mutual repulsion of water and greasy ink. Damping rollers pass over the surface followed by inking rollers. The image area, which is greasy, repels the water but retains the ink (also greasy), which is transferred to the paper. In modern practice, a sheet of zinc or aluminium, grained to retain moisture, is used in place of stone. A sheet of paper coated with a hydrolytic surface can also be used for printing purposes.

LOWER CASE – Small letters of the alphabet.

MAKE READY – Necessary operations to prepare a duplicating machine or printing press for operational purposes.

MASK – Opaque material used to prevent parts of the original from being registered on the photo-receptor.

MASTER – A document, in the form suitable to the reprographic process being used, from which copies can be made. A master is specially prepared but in some processes may be the original.

MASTER MAKING – A term used to describe the activity associated with the production by any means of an intermediate or a master to be used reprographically.

MAT – A colloquial term used to describe a document specially treated to receive a typed or drawn image for reproduction by the offset lithographic process. It is also used to describe a paper offset printing plate.

MATRIX – An arrangement of individual items in a regular or predetermined form.

MICROCOPY – A copy obtained by means of an optical device whereby the copy has dimensions greatly reduced in comparison with the dimensions of the original document.

MICROFICHE – A transparent, flat, rectangular sheet of photographic film bearing a number of micro-images arranged in horizontal and vertical rows. The microfiche also normally displays the title of the copied document in a size large enough to be eye-legible.

MICROFILM – The photographic film used in micro-copying; to produce micro copies.

MICROFORM – A generic term for all types of micro copy.

MICRO-IMAGE – An image obtained by means of an optical device whereby the image has dimensions greatly reduced in comparison with the dimensions of the original.

NEGATIVE – Having a tonal arrangement opposite to that of the original. Generally used to refer to copies which have a white image on a black background. Also used to describe an electric charge or state where there is an excess of electrons.

NUCLEUS – An atomic nucleus is composed of protons and neutrons and constitutes practically all the mass of the atom. Its charge equals the atomic number; its diameter is from 10^{-13} to 10^{-12} cm.

OFFSET LITHOGRAPHY – A printing process in which the inked image is formed lithographically but is transferred from the master to a blanket from which it is offset on to the medium which carries the message and/or design.

OFFSET PRINTING – The most commonly used form of lithography in which the image is transferred from the master to the final material via an intermediate surface.

OPAQUE – Having such a resistance to the passage of light as to be unsuitable for processes using transmitted light.

OPTICAL COPYING – Producing copies by means of an optical system.

OPTICAL PRINTING – Printing copies by means of equipment incorporating an optical system to create the image.

ORIGINAL – The document that is to be copied, duplicated or printed.

OVERLAY(S) – A medium consisting of opaque and/or transparent areas which allow parts of an original document selectively to be masked, changed or copied according to the design of the overlay and used for electrostatic reprographic paperwork routines and systems.

PAPER PLATE – A term used to describe a direct image offset litho master.

PASTE-UP – An original created by pasting selected parts of varying subject matter into a desired format.

PCMI – Photochromic Micro Image.

PHOTOCHROMIC MICRO IMAGE – A trade name of the National Cash Register Company for a high-reduction (above 1:150) micro-copying process employing substances which change their colour when subjected to luminous radiation.

PHOTOCOPY – A copy produced on or by means of material sensitive to radiation.

PHOTOGRAVURE – A printing process which uses an etched plate produced by photographic means in which the etched areas represent the image. The tonal intensity of the image produced is controlled by the depth of the etch, the deeper the etch the greater amount of ink is held and ultimately transferred to the medium being printed.

PHOTOLETTERING – Preparing text by contact or optical printing on light sensitive material.

PHOTOLITHO – A process for producing lithographic or offset lithographic printing masters by photographic means.

PHOTOSTAT – A term derived from the original trade mark owned by the Photostat Corporation now in common use to describe a right reading negative photocopy, or positive prints made from it produced on a Photostat or similar camera.

PHOTO-RECEPTOR – A surface, curved, flat, rigid or flexible designed to receive and retain visual images.

PICA – A large printing type; a standard of printing measurement (12 points).

PICA-EM – Unit of measurement representing 12 points or approximately $\frac{1}{6}$ in.

POINT – Printing unit of measurement, approximately $\frac{1}{72}$ in.

POLARITY – The distinction between the north and south poles of an electric machine or circuit, usually described as negative or positive.

POSITIVE – Having the same tonal arrangement as those of the original. Generally used to refer to copies which have a black image on a white background. Also used to describe an electric charge or state which is deficient of electrons.

POSITRON – A sub-atomic particle of mass and charge equal to those of the electron, but having its charge positive.

PRINT – A reproduction of an original document, but usually a reproduction of a master made from the original document; to make reproductions of original documents directly, or indirectly by means of an intermediate or a special master.

PRINT ROOM – A place where reprographic equipment is located and operated.

PROCESSING – This includes all the steps and treatments that take place after an original is created and the final copies are produced including their finishing.

PROPORTIONAL SPACING – Allocation of space in the horizontal direction appropriate to the width of the character.

PROTON – A positively charged particle of mass $1 \cdot 66 \times 10^{-24}$ gm and a charge equal and opposite to that of an electron. The proton being 1840 times as massive as the electron, it makes the main contribution to the mass of an atom.

REAM – 500 sheets of paper of one size.

REDUCTION – Making to a smaller scale than 1:1

REFLEX COPYING – Contact copying when radiation first passes through the copy support and is then reflected on to it by the original.

REGISTER – Exact alignment of type, especially when overlaying one image upon another.

REPROGRAPHY – The technology of producing and reproducing two dimensional visual image media in business and administrative operations.

REPROTYPING – The preparation of text on originals or masters in a way suitable for copying, duplicating or printing by any reprographic method.

RESOLUTION – A measure of the sharpness or clarity of an image.

REVERSAL DEVELOPMENT – A method of development that produces a positive from a positive or negative from negative.

REVERSE READING – Having the image left to right as in a mirror; this effect is also called mirror reading.

RIGHT READING – An image which is directly readable, as opposed to a laterally reversed image.

ROTARY COPIER – This term is usually associated with electrostatic copiers and duplicators in which the original enters the machine, is retained in it and rotates over a drum or similar device during the copying or duplicating cycle, after which it emerges from the machine with the copies created from it.

RUN – The number of copies produced from each master or each original when the original is used as the master.

SCORATRON – An electrical device with a screen for creating a corona discharge.

SCRAPER BLADE – Another name for a doctor blade.

SCUM – A thin coating of ink which appears on litho and offset litho masters due to a breakdown in the desensitizing solution.

SELENIUM – A non-metallic element in the sixth group of the periodic system. The dark (metallic) allotropic form created when other forms are heated at 200°C. is a conductor of electricity when illuminated and it is in this form that it is used with additives in xerographic copiers and duplicators.

SENSITIZED – The treatment of materials used in reprography so that they become sensitive to radiation which is normally light or heat.

SHARPNESS – The quality of an image, characterised in particular by well marked contours which prevent confusion or ambiguity in the distinguishing of details.

SHELF LIFE – In reprography the period of time from the manufacture of materials to the time they can be used with acceptable results provided that they have been stored unopened in recommended conditions.

SHINGLING – A method of laying out documents, usually cards, for reproduction purposes so that only one edge, generally the top edge, shows one or two lines of typescript while the rest of the subsequent documents or cards mask the remainder of the document or card beneath the part that remains visible.

SHUTTER – A device in a camera or other optical instrument to shut off or to allow through luminous radiation during a known time or at will.

SOFT CENTRE EFFECT – This describes those areas of an original which have been rendered on copies of it as a less dense tone than the edge of the area of the image.

SPECTRAL RESPONSE – The comparative response of a photo-receptor to exposures of light of different wavelengths but of constant intensity.

SPIRIT DUPLICATING – A process which transfers the impression directly from a reverse reading master carrying an aniline dye image to the copy paper by means of a fluid and pressure.

STENCIL DUPLICATING – A process employing a specially prepared master whose image is imposed in the form of cuts made in an ink-impervious coating supported by a strong tissue base. Special ink is forced through the coating and transferred to the copy material by pressure.

STOCK – A printers term for paper.

STOP – Alternative name for aperture.

STRIP INDEX PANEL – A rigid or flexible panel designed to accommodate strips of paper, card stock or other semi-flexible media which carries only a few lines of data and information so that this can be moved about the panel at will and the order in which the information appears can be changed at will.

TEMPLATE – Ancillary device used with electrostatic copiers and duplicators to accommodate originals in various ways so that they can be processed in accordance with specially designed paperwork systems and routines. Thus, there are shingling templates, masking templates, four-up templates, titling templates which are used to alter or delete information from documents electrostatically copied or duplicated.

THERMOGRAPHY – A direct or indirect process for producing copies on special material which uses heat for the formation of the image.

THERMOPLASTIC – Becoming plastic when heated. Specifically any resin which can be melted by heat and then coled. The process being able to be repeated any number of times without appreciable change in the properties of the material.

THERMOSETTING – The action of moulding plastic compounds under heat and/or pressure, inducing an irreversible chemical change.

THROAT – The aperture of a copying or duplicating machine into which the original has to be inserted for reprographic processing.

TONER – The finely divided pigmented solid particles used to develop latent electrostatic images.

TRANSFER LETTERING – Printed characters set on a translucent carrier in such a way as to facilitate the pressure transfer of selected characters and designs to art work.

TRANSLUCENT – Permitting the passage of light, the emergent light being diffused.

TRANSPARENCY – An original or master designed for viewing, projection or for copying purposes by means of light being transmitted through it. The emergent light does not become diffused.

TRIBO-ELECTRIFICATION – The electrical polarization of non-conductive materials by friction.

TYPOGRAPHY – The art and technique of arranging printing. The appearance, arrangement and style of reading material.

ULTRA-VIOLET – Invisible radiation having a shorter wave length than that of visible light.

UNIT-SPACING – Allocation of equal space in the horizontal field to each character irrespective of character width.

UNITIZED APERTURE CARD – A card, usually the same size as a punched card, with a square aperture to accommodate a piece of film usually of 35 mm wide.

UPPER CASE – Capital letters.

VISUAL COMMUNICATION – Any type of communication that can be transmitted to the brain directly by the visual sense.

XEROGRAPHY – A process for making copies of printed, written or pictorial material from film or paper, in which areas on a photo-receptor corresponding to those areas on the original that are to be reproduced are sensitized with a charge of static electricity so that, when powdered with a coloured resin carrying an opposite charge, only the charged areas retain the resin, which is then fused to the photo-receptor or transferred to plain paper and fused to make it permanent. Pronounced "zee-ro-gra-fee".

INDEX

Aerosol development 77
Allwachs W. H. 43
A. M. Corporation 128
  A. M. machines 60, 102, 104, 132, 133
Apeco machines 102
Application(s) 9, 84
  Limitations 9
  Systems 12, 86
  Templatry 11, 86
Arcor machines 104
Atomic physics 44

Battelle Memorial Institute 23
Binders 50
Bixby 24
Bonding the image 34, 36, 37, 38, 68, 69

Canon NP System 54, 133
Coratron 45
Corona discharge 45
Carlson, Chester F. 19
Cascade developement 72
Colour printing 127
Competitive processes 135
Convenience factors 123
Copycat machines 106
Copymate 136
Cost per copy calculations 101
  Check list 118

Dalcopy machines 109
De Meuleneare 19
Dennison Manufacturing Company 128
  Dennison copiers 106
Dessauer Dr. John 23
Developement 33, 68
  Aerosol 77
  Cascade 72, 131
  Colour 129
  Control 78
  Disposal 75
  Fur brush 75
  Liquid 33, 74, 75, 129
  Magnetic brush 75, 131
  Methods 72 - 77
  Powder cloud 77
  Replenishment 72, 75
  Reversal 78
Dick A. B., Company 27
  A. B. Dick Machines 38, 65, 108
Diazo (or Dye-line) process 17

Edge effect 54, 74, 75
Elfasol 29, 35, 36, 52, 74
Electrofax 15, 25, 29, 35, 36, 40, 129, 130
Electrography 18, 27, 43
  Processes 38
  Machines 12
Electron(s) 44
Electrophoresis 74
Electrophotography 10, 18, 21
  First image 21
  First steps 18
Electrostatic(s) 10, 17, 27, 43
  Commercial processes 12, 29
  3D Printing 148
  How the processes work 27
  Machine specifications 102, 117
  Printing 9, 10
  Process capability 10
  Process limitations 14
Equipment characteristics 102, 107
  Check list 102, 103

Ferro-prussiate process 17
Fixing agents 75
Flat-bed machines 12, 13
Furnium 53
Fusing the image 33, 34, 36, 37, 69

Gevafex machine 104

Haloid Company 23
Huebner William C., 77

Illumination of original 58, 63

Image 56
  Path 60, 62
  Development 68
Ionic transfer 72
Ionisation 46

Kalle process 28, 31, 36, 96
Kornei, Otto 21

Lens function of 56, 58
  Adjustments 59
Lens-less equipment 56, 64
Letterpress 9
Limitations 14, 102 to 117
Long distance xerography (LDX) 137, 139

Managerial objectives 97
Map making 129
Masks 11, 12
Master 10
May 43
Metcalfe 25
Microfiche 140
Microfilm 140
Microform 140
Minnesota Mining and Manufacturing Company 25, 127, 136
Minolta machines 110

Neutron(s) 44
Nucleus 44

Offset lithography 9
Optical Society of America 24
Ordoverax Veleography 17
Originals 80
  Analysis of 123
  Characteristics 84
  Check list 82
  Examination 82
  Handling 80
  Improvement of 84
  Preparation 86, 87
  Types of 81
Overlays 11, 12
Ozalid Limited 17

Paper 89
  Characteristics 93, 94
  Effect on machine design 95
  Grain 93
  How made 91
  Storage and use 92
Persistent internal polarization (PIP) 133, 134
Phol R. 18
Photo-conductors 43, 47
  First noted 43
  Purity 50
Photo-electrophoresis 128, 131
Photography 17
Photon 58
Photo-receptor(s) 31, 35, 43, 130, 134
  Additives 31
  Cadmium sulphide 54
  Carriers 31
  Characteristics 50
  Charging of 45
  Discharging of 47
  Electrical qualities 43
  How they work 43
  Imaging 56
  Types 51, 52, 53, 54
Pitney-Bowes machine 110
Powder cloud development 77
Pressureless printing 9, 148
Printing 9
Process 27
  Choice of 99
  Projections 127
Proton 44

Radio Corporation of America (R.C.A.) 25
Rank Xerox machines 61, 62, 114
RCA Review 130
Remington Rand machine 112
Reprographic costings 120
Reprography 17, 20, 80
  Centralised operation 15
Review of equipment 101
Rotary machines 12

Schaffert R. M., 24, 28, 133
SCM. Corporation 128
  SCM copiers 60, 116
Scoratron 45
Selenyi, P. 18
Semi-conductors 43
Shingling 13, 86
Smoke printing 77
Soft centre effect 54
Statikon machine 116
St. Elmo's fire 46
Strip index 13
Systems applications 10

Telecopier 137

Tellurium 30
Templates 11, 13
Toner 27, 33, 35
  Characteristics 68, 69
  Charging of 70
  Coloured toners 69
  Storage of 70
Tribo electrification 71

Videofile 139, 140, 141
Videograph 18, 38, 137
Video jet process 40, 137
Von Bronk 18

Walkup and Wise 24

Xerography 22, 34, 36
  Definition 28
  Derivation of name 24
Xerox Corporation 29, 128

Zinc oxide 51, 52